GYROSCOPES

TYPES, FUNCTIONS AND APPLICATIONS

MECHANICAL ENGINEERING THEORY AND APPLICATIONS

Additional books and e-books in this series can be found
on Nova's website under the Series tab.

GYROSCOPES

TYPES, FUNCTIONS AND APPLICATIONS

MARCEL GERSTE
EDITOR

nova
science publishers
New York

NOTICE TO THE READER

Additional color graphics may be available in the e-book version of this book.

Library of Congress Cataloging-in-Publication Data

ISBN: 978-1-53615-856-4

Published by Nova Science Publishers, Inc. † New York

CONTENTS

PREFACE

In the first chapter of Gyroscopes: Types, Functions and Applications, the operating principle, types and applications of fiber-optic gyroscopes are summarized, and a novel slow light gyroscope based on coupled resonators is introduced.

Following this, two signal processing techniques are discussed, each for a different type of gyroscope noise. These two techniques are then combined to produce a general technique for improving the accuracy of a gyroscope.

The main methods of creating compact passive optical gyroscopes and their development trends are examined in the closing chapter.

Chapter 1 - The gyroscope, a device which is used to measure angular velocity and angular displacement has experienced the development of the first generation of an electromechanical gyroscope and the second generation of a laser gyroscope. With the development of fiber-optic technology, the third generation of the fiber-optic gyroscope (FOG) based on the Sagnac effect has been born. Compared with the traditional gyroscope, the FOG has an all-solid structure with no moving and wearing parts, as well as having no self-locking effect. With the advantages of low cost, long life, and anti-electromagnetic interference, etc, it can be widely used in military and civil fields. According to its operating principle, FOG can be divided into three types: interference gyro (I-FOG), resonance gyro

(R-FOG) and stimulated Brillouin scattering gyro (B-FOG). In this chapter, the operating principle, types and applications of FOGs are summarized, and a novel slow-light gyroscope based on coupled resonators is also introduced.

Chapter 2 - For many systems requiring angular velocities of a distributed mass, such as guidance and control systems for unmanned space vehicles, it is highly desirable to have inertial measurement sensors that are small, inexpensive, low power, reliable and accurate. Technological advances in the design and construction of micro inertial sensors, such as accelerometers and gyroscopes, have much promise in providing small, inexpensive, and low power devices; however, additional improvement in the reliability and, especially, the accuracy of these micro devices is still necessary. Although major improvements in these two properties may occur in the future, in this chapter it is proposed that signal processing methods be used to provide appropriate accuracies and, in many cases, improved reliability. Specifically, it is proposed that appropriate gyroscope accuracy can be attained by using statistical methods to combine the output measurement of many, perhaps one hundred or more, MEMS gyrocopes on a single chip (or a few chips) to provide a single accurate measurement. One method of performing such a combination is through an extended Kalman filter (EKF). A standard application of an EKF to an array of gyroscopes would involve at least six state equations per gyroscope and the number of covariance equations would be in the order of the square of the product of six times the number of gyroscopes. Obviously, the 'curse of dimensionality' produces an explosion of computations. Even if the EKF for each individual gyroscope is uncoupled from the rest, the number of covariance equations is of the order of the number of gyroscopes times six squared, which can still lead to a formidable computational burden. In this chapter, two signal processing techniques are discussed, each for a different type of gyroscope noise. These two techniques are then combined to produce a general technique for improving the accuracy of a gyroscope. The final gyroscope output is generated by an EKF with a total of six state equations and (no more than) thirty six covariance equations.

Chapter 3 - Currently, optical gyroscopes are widely used in inertial navigation systems. At the same time, there is a pronounced tendency towards miniaturization, which spreads also to the navigation systems. The modern economics requires the development of mini and micro sensor devices to control highly dynamic technical systems, mobile robots, microsatellites, etc. The size of the controlled objects is continuously reduced, which in turn requires further miniaturization of gyroscopes. In response to this challenge during the last decades two directions in reducing the dimensions of optical gyroscopes were developed: one of which is aimed at creating compact active gyroscopes, and the second – compact passive gyroscopes. This chapter is devoted to the analysis of the main ways of creating and the development trends of compact passive optical gyroscopes. Their sizes are comparable to micromechanical (MEMS) gyroscopes sizes, and their ultimate sensitivity significantly exceeds the sensitivity of micromechanics and approaches the sensitivity of laser and fiber gyroscopes. Often such gyroscopes are called micro-optical (MOG) based on the size of their basic elements. Research on this topic was pushed forward by the development of integrated optics, and most of the prototypes of compact passive optical gyroscopes are built on the basis of optical integrated circuits, so these gyros are frequently called integrated optical gyros. Although the Sagnac effect is the basis of the operation of all optical gyroscopes, the method for determining the angular velocity depends on the connection circuit of their sensitive element. In the framework of this chapter compact passive optical gyroscopes are considered, first of all, from the point of its sensitive element connection circuit. In addition, various possible types of MOG sensitive elements are discussed since the characteristics of gyroscopes and optimal design are associated with their choice.

In: Gyroscopes
Editor: Marcel Gerste

ISBN: 978-1-53615-856-4
© 2019 Nova Science Publishers, Inc.

Chapter 1

THE TYPES, FUNCTIONS AND APPLICATIONS OF FIBER-OPTIC GYROSCOPES

Yundong Zhang[1],*, Weiguo Jiang[1], Jinfang Wang[2], Yatong Zhang[1] and Yaming Chen[1]
[1]National Key Laboratory of Tunable Laser Technology, Institute of Opto-Electronics,Harbin Institute of Technology, Harbin, China
[2]Shanghai Engineering Research Center of Inertia, Shanghai Aerospace Control Technology Institute, Shanghai, China

ABSTRACT

The gyroscope, a device which is used to measure angular velocity and angular displacement has experienced the development of the first generation of an electromechanical gyroscope and the second generation of a laser gyroscope. With the development of fiber-optic technology, the

---

* Corresponding Author's E-mail: ydzhang@hit.edu.cn.

third generation of the fiber-optic gyroscope (FOG) based on the Sagnac effect has been born. Compared with the traditional gyroscope, the FOG has an all-solid structure with no moving and wearing parts, as well as having no self-locking effect. With the advantages of low cost, long life, and anti-electromagnetic interference, etc, it can be widely used in military and civil fields. According to its operating principle, FOG can be divided into three types: interference gyro (I-FOG), resonance gyro (R-FOG) and stimulated Brillouin scattering gyro (B-FOG). In this chapter, the operating principle, types and applications of FOGs are summarized, and a novel slow-light gyroscope based on coupled resonators is also introduced.

Keywords: gyroscope, optical fiber, Sagnac effect, rotation sensor, I-FOG, R-FOG, B-FOG, slow-light, coupled resonators

1. INTRODUCTION

The gyroscope is a sensor that measures the rotation angle or angular velocity of a moving object in inertial space. The development of the gyroscope has been a long and arduous period, and has made great achievements up to today. In 1852, French physicist J. Foucault found that the direction of a rigid body is stable under high speed rotation. He borrowed this phenomenon and developed the world's first gyroscope based on mechanical rotation. Later, German engineer H. Anschutz and American engineer Sperry designed the gyro compass in 1908 and 1909 respectively, which were used for navigation by ships at sea. In 1913, French scientist G. Sagnac demonstrated that optical systems with no moving parts can also detect rotation in relative inertial space.

In the 1960s, with the development of laser technology, optical gyroscopes based on the Sagnac effect began to develop rapidly. In 1962, Heer C. V. and Rosenthal A. H. proposed the concept of a ring laser gyroscope, and developed the world's first ring laser gyroscope device in 1963. However, the ring laser gyroscope is not an all-solid state device since it uses mechanical dithering excursion of cantilever beam structure to avoid locking.

After the 1970s, with the increasing maturity of optical fiber technology, the fiber-optic gyroscope (FOG) with all-solid structure emerged as the times required. In 1976, Vali V and Shorthill R W from the University of Utah first put forward the idea of FOG and carried out experiments [1]. As a new type of all solid state inertial instrument, FOG has many advantages that traditional electromechanical gyroscopes or ring laser gyroscopes lack, which can be summarized as follows:

1. *High sensitivity.* FOG can enhance the Sagnac effect in fiber coil by appropriate methods which greatly improves the sensitivity of Sagnac phase shift detection.
2. *All-solid structure.* Compared with the traditional electromechanical gyroscope, the structure of FOG has no rotating and wearing parts, as all parts are fixed on the rigid framework of the gyroscope. FOG has the advantages of small size, light weight, long life, and a simple production and assembly process.
3. *Flexible structural design.* FOG can adapt to the requirements of different dynamic ranges and accuracy levels without changing the structure and circuit components of the gyroscope, only by changing the parameters of the optical path. It can cover all the ranges of the gyroscope in aviation, aerospace, navigation and land applications.

2. PRINCIPLE

2.1. Theory of the Sagnac Effect

In 1913, the French physicist G. Sagnac demonstrated that the rotation of an object's relative inertial space could be measured by an optical system without moving parts. This theory is well known as the Sagnac effect. The classic Sagnac optical interferometer is an extension of the Michelson-Morley interferometer, and its schematic diagram is shown in Figure 2.1. The beam emitted from the light source is divided into two beams by the semi-permeable reflector BS (M_0), respectively, forming a clockwise beam path L_{cw} and a counter-clockwise beam path L_{ccw}. The beam along Lcw is reflected by the reflector M_1, M_3 and M_2 successively, while the beam along the Lccw is reflected by the reflector M_2, M_3 and M_1 successively. Then the two beams merge and interfere through BS, and the intensity is detected by the detector. When the whole system is stationary, $L_{cw} = L_{ccw}$, and the center of the interference fringe is bright. Once the system rotates, $L_{cw} \neq L_{ccw}$, the interference fringes move, so the rotation of the system can be detected.

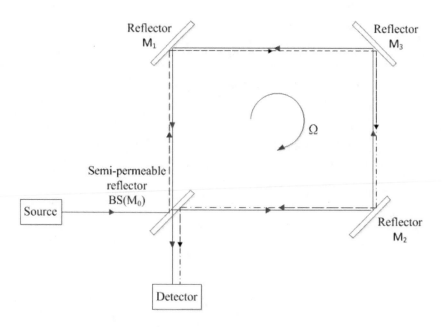

Figure 2.1. Sagnac interferometer composed of discrete components.

In the actual fiber-optic gyroscope, the two back-propagation optical paths are not set up of discrete optical elements, but are composed of fiber coils, so the Sagnac effect in optical fiber is derived as below. Figure 2.2 demonstrates the Sagnac effect in fiber loop. Assuming that the radius of the fiber loop is R, two beams emitted from arbitrary point M, one beam propagates clockwise (CW) and the other propagates counterclockwise (CCW). When the fiber loop is stationary, the two back-propagation beams still converge at point M after propagating around for one cycle and overlap at point M to form interference. The transmission times are $t_{cw} = t_{ccw} = N \cdot 2\pi R/c$, where N and c are the loop number and the speed of light in vacuum respectively. The propagation velocity and the distance propagated are all equal, which illustrates that the optical path difference is zero, so that there's no phase difference.

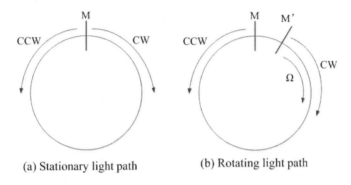

(a) Stationary light path   (b) Rotating light path

Figure 2.2. Sagnac interferometer in optical loop.

When the loop rotates with the angular velocity Ω (assuming that it rotates clockwise), and after the beam propagates one cycle where the loop has turned an angle, the starting point of the beam changes from M to M'. The distances propagated through CW and CCW are not equal and the optical path's changes will cause phase changes. Once the phase difference between the two beams is detected, the rotation angular velocity Ω of the loop can be obtained through a certain calculation. The starting point of beam from M to M' can be understood in two ways: One way is that the velocity of the beam does not change, and CW takes a longer time than CCW; the other is that the propagation path of the beam remains unchanged, and the velocity of CW beam decreases while that of CCW beam increases. This chapter adopts the latter to deduce the relationship between the phase difference ϕs and angular velocity Ω. When the loop rotates, the velocities of CW and CCW are not equal. According to the velocity synthesis formula of classical mechanics, the velocities of CW beam and CCW beam are respectively:

$$c_{cw} = c - R\Omega \tag{2.1}$$

$$c_{ccw} = c + R\Omega \tag{2.2}$$

The corresponding propagation times are:

$$t_{cw} = \frac{N \cdot 2\pi R}{c_{cw}} = \frac{N \cdot 2\pi R}{c - R\Omega} \tag{2.3}$$

$$t_{ccw} = \frac{N \cdot 2\pi R}{c_{ccw}} = \frac{N \cdot 2\pi R}{c + R\Omega} \tag{2.4}$$

Due to $c^2 \gg (R\Omega)^2$

The time difference between CW and CCW beams is

$$\Delta t = t_{cw} - t_{ccw} = N \cdot 2\pi R \cdot \frac{2R\Omega}{c^2 - (R\Omega)^2} \approx \frac{N \cdot 4\pi R^2 \Omega}{c^2} \tag{2.5}$$

So, the phase difference between CW and CCW beams is

$$\phi_s = \frac{2\pi c}{\lambda} \cdot \Delta t = \frac{2\pi c}{\lambda} \cdot \frac{N \cdot 4\pi R^2 \Omega}{c^2} = \frac{8\pi A}{\lambda c}\Omega = \frac{4\pi RL}{\lambda c}\Omega \tag{2.6}$$

where λ is the wavelength of the beam, $A = N\pi R^2$ is the total area of the closed loop, $L = N \cdot 2\pi R$ is the total length of the loop.

One can see from the above equation that for a certain gyroscope, if the beam wavelength λ is stable, the phase difference ϕ_s is proportional to the angular velocity Ω. The value of Ω can be calculated if ϕ_s is known. In general, when λ is stable, the increase of the radius R or the length L can make the value of ϕ_s larger. The sensitivity of the gyroscope can be improved in this way.

2.2. Main Performance Index of the Fiber-Optic Gyroscope

The performance of the FOG is mainly evaluated by the following indexes:

2.2.1. Scale Factor

Scale factor is the ratio of the output value to the input angular velocity, reflecting the sensitivity of the gyroscope. Scale factor is usually expressed in the slope of a particular line. The line can be fitted by the least square method from the output/input data by changing the input angular velocity within the scope of the entire input. Scale factor stability is a comprehensive index to measure scale factors, including scale factor nonlinearity, scale factor asymmetry and scale factor repeatability. The square root of the above three errors is called scale factor stability.

There are many factors that affect scale factor stability, such as the instability of fiber polarization, the change of the fiber loop, the instability of excitation signal period, the change of ambient temperature and the modulation/demodulation scheme.

2.2.2. Zero-Bias Stability

In FOG, drift and noise are two different concepts. Zero-bias stability, namely drift, refers to the fluctuation of gyroscope output around its mean value, which is generally expressed by standard deviation (σ) or root-mean-square difference (RMS). Therefore, it implies that the probability properties of the above random processes are normally distributed. In this sense, the drift value also indicates the dispersion degree of the observed value around the zero deviation mean. For fast response applications (short term), zero-bias stability is determined by noise; while for navigation applications (long term), zero-bias stability is mainly due to low frequency perturbations that change slowly.

2.2.3. Random Walk Coefficient

The noise of FOG is mainly generated from the photoelectric detection part, which determines the minimum detectable sensitivity of FOG. These noises include polarization, backscatter and Rayleigh scattering noise, as well as shot noise of the detector, which are called white noise. Since the zero-bias stability is related to the detection bandwidth, the larger the detection bandwidth is, the smaller the zero-bias stability will be. Therefore, it is difficult for this index to describe the white noise of FOG. Therefore, the random walk coefficient is introduced to represent the size of white noise. It is defined as the ratio of the measured stability of zero-bias to the square root of the detected bandwidth, its unit is $((°) \cdot h^{-1})/\sqrt{Hz}$ or $(°)/\sqrt{h}$..

2.2.4. Dynamic Range

The dynamic range is the ratio of the maximum input angular velocity to the minimum detectable angular velocity. For I-FOG, the response is a cosine function, there is a zero centered $\pm\pi$ radian monotone phase measurement interval. The rotating velocity also has a corresponding single-value working range of $\pm\Omega_\pi$:

$$\Omega_{max}=\Omega_\pi=\frac{180}{\pi} \cdot \frac{\lambda c}{4LR}((°)/s)$$

(2.7)

where λ is the wavelength of the incident light, c is the speed of light in vacuum respectively, L and R are the total length and radius of the loop.

With the increase of length and radius of the FOG, its sensitivity is higher, but the corresponding Ω_π is smaller, so that the dynamic range remains unchanged.

2.3. Reciprocity of Optical Path

Assume that the precision is 0.01°/h - 0.001°/h, the length of the loop is 1000 m, the radius of the ring is 0.1m, and the wavelength of the beam is 1.3×10^{-6} m. According to Formula(2-6), the corresponding Sagnac phase shift is 10^{-7}rad-10^{-8}rad, and the optical path difference between clockwise and counterclockwise is 10^{-14}-10^{-15}m. This small phase difference is drowned out by the cumulative phase change along the propagation direction. To solve this problem, it is necessary to make the two beams of backpropagation have the same transmission characteristics, and the reciprocal structure can make the additional phase shift of the two beams of backpropagation be the same, and eliminate various parasitic effects.

This problem can be solved with a so-called minimum reciprocity structure (Figure 2.3) where light is fed into the interferometer through a single-mode waveguide and where the returning light wave is filtered through this same waveguide at the output. [2] This ensures that both returning waves have experienced exactly the same total phase despite potential defects in the splitting. The reciprocal operation does not require a continuous single-mode propagation but merely a single-mode filtering at the common input-output port called the reciprocal port. So, it greatly reduces the cost, and is more conducive to practical applications.

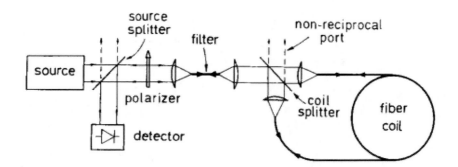

Figure 2.3. Minimum reciprocity structure.

3. TYPES OF FIBER-OPTIC GYROSCOPE

According to the working principle, FOG can be divided into three types: Interferometric fiber-optic gyroscope (I-FOG), Resonant fiber-optic gyroscope (R-FOG) and Stimulated brillouin scattering fiber-optic gyroscope (B-FOG). Among them, the I-FOG was developed earliest. At present, the I-FOG products have been widely used in various carriers of sea, land, air and space, which can meet the precision requirements in civil and military. The other two types of gyroscopes were developed relatively late and are still in the stage of theoretical research and experimental verification.

3.1. Interferometric Fiber-Optic Gyroscope

The main body of the I-FOG is a Sagnac interferometer, which consists of a light source, couplers, a photodetector, and optical fiber coils. The principle is based on the Sagnac effect. The schematic diagram of a typical I-FOG is demonstrated in Figure 3.1. The two couplers are both 3dB couplers, which ensure the reciprocity of the optical path. The beam emitted from the coupler C_2 is divided into two beams, which pass through clockwise and counterclockwise respectively, and then converge in C_2, resulting in interference. According to the derivation in the previous section, when the I-FOG rotates, a phase difference ϕ_s, which is proportional to the rotation angular velocity Ω, will be generated between two beams propagating clockwise and counterclockwise in the fiber coil:

$$\phi_s = \frac{4\pi RL}{\lambda c} \cdot \Omega \tag{3.1}$$

Since ϕ_s is proportional to the radius and length of the fiber coil, different structures can be proposed to meet the application requirements in different occasions under the condition of the same overall scheme. It is the flexibility in design that distinguishes FOGs from electromechanical and laser gyroscopes. According to the different applications, I-FOG can be roughly divided into four types: rate level, tactical level, inertial level and precision level. The technical indicators are shown in Table 3.1:

Table 3.1. Accuracy level and technical requirements of I-FOG

Level	Zero-bias stability((°)/h)	Scale factor stability
Precision level	<0.001	<1 ppm
Inertial level	0.01	<5 ppm
Tactical level	0.1-10	10-1000 ppm
Rate level	10-1000	0.1%-1%

3.1.1. Principle of I-FOG

Since the photodetector directly detects the intensity of light waves, the phase information contained in it must be demodulated by the intensity signal of the detector, and then the velocity of rotation can be obtained by Formula (2-1). Light amplitude on the detector is the sum amplitude of the clockwise and counterclockwise waves of Sagnac interferometer, and there exists a phase shift ϕ_s between the two backpropagation waves due to rotating. As shown in the I-FOG structure of Figure 2.3, we can get:

$$E_{out} = E_{cw} e^{i\phi_s} + E_{ccw} \tag{3.2}$$

Assuming that each coupler has an accurate 1:1 spectroscopic ratio, so that $E_{cw} = E_{ccw} = E_0$, the interference photosynthetic amplitude is:

$$E_{out} = E_0(1+e^{i\phi_s}) = E_0(1+\cos\phi_s + i\sin\phi_s) \tag{3.3}$$

where E_0 is the amplitude of light wave, incident to Sagnac interferometer of each direction.

Then the intensity of the interference light wave returning to the detector is:

$$I_D = \frac{1}{2}E_{out}^2 = \frac{1}{2}E_0^2[(1+\cos\phi_s)^2 + \sin^2\phi_s] = I_0(1+\cos\phi_s) \tag{3.4}$$

where, $I_0 = E_0^2$.

The relationship between the output light intensity of the FOG and the Sagnac phase shift is shown in Figure 3-1. Since I_D is the cosine function of ϕ_s, one can see from the curve that:

1. When the Sagnac phase shift is small, the sensitivity of the gyroscope output to rotation rate is close to zero.
2. Due to the symmetry of the response curve, it is not possible to determine the sign of the Sagnac phase shift or the direction of rotation.
3. The response curve is periodic, and the gyroscope output is multivalued when measuring ϕ_s over two or more periods, which can be a problem especially when starting the gyroscope at high speed. In order to uniquely determine the magnitude of the rotation angular rate, it usually requires the Sagnac phase shift ϕ_s within the first interference fringe.

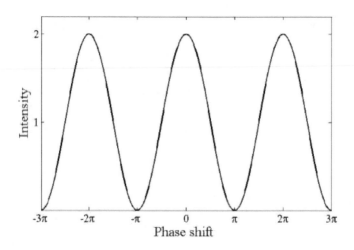

Figure 3.1. The curve of the relationship between the output light intensity of the I-FOG and the sagnac phase shift.

3.1.2. Photonic Crystal Fiber-Optic Gyroscope

Until now, the development of I-FOG has been quite mature. In the world, medium and low precision I-FOG has been widely used in aerospace, navigation, weapons systems and other industrial fields. At present, the research of I-FOG is mainly focused on miniaturization, integration and precision. Among these, high precision is meant to propose various methods to eliminate or reduce the various non-reciprocity errors in gyroscopy. Scientists put forward many methods to improve the accuracy of I-FOG. This chapter focuses on the most popular photonic crystal fiber optic gyroscope in recent years.

As a new type of micro-air pore structure, photonic crystal fiber (PCF), compared with traditional single-mode or polarization-maintaining fiber, has unmatched superior characteristics, such as high birefringence, infinite single-mode and flexible dispersion controllable characteristics. Due to its stable temperature characteristics, weak magnetic sensitivity and low bending loss, PCF shows great potential in the field the of fiber-optic gyroscope. So the photonic crystal fiber-optic gyroscope has been widely studied to improve the accuracy of I-FOG.

In 2006, Draper Laboratory of Cambridge, Massachusetts in the United States [3], developed a photonic crystal fiber-optic gyroscope with accuracy better than $0.02\,(°)/h$. Its structure is shown in Figure 3.2. The gyroscope adopts a solid-core photonic crystal fiber, which has the characteristics of infinite single mode, and the relative intensity noise can be effectively suppressed by using a broad spectrum light source. The solid-core photonic crystal fiber also has the characteristics of smaller diameter and bending radius, which can effectively reduce the volume of the fiber loop.

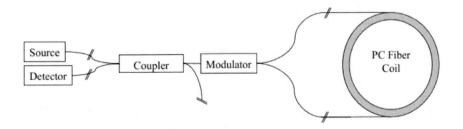

Figure 3.2. FOG based on solid-core photonic crystal fiber proposed by Draper Laboratory.

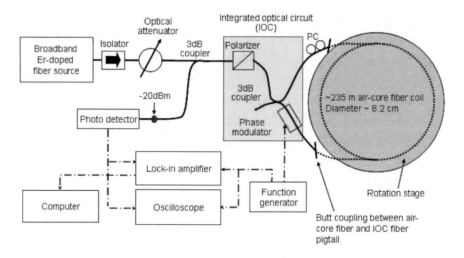

Figure 3.3. FOG based on air-core photonic crystal fiber proposed by Ginzton Laboratory.

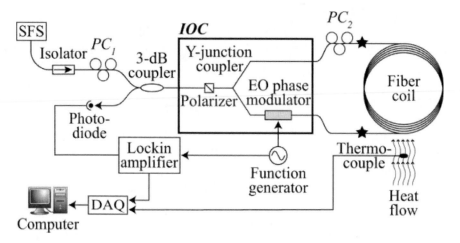

Figure 3.4. Air-core photonic crystal fiber gyroscope proposed by Ginzton Laboratory.

In 2006, Ginzton Laboratory of Stanford University proposed the first air-core photonic crystal fiber FOG [4], whose diagram is shown in Figure 3.3. Its zero bias stability is about 2 (°)/h, and the minimum detectable angular rate is 2.7 (°)/h. From the perspective of long-term stability, the drift introduced by the Kerr effect in the air-core photonic crystal fiber FOG is about 1/170 of the traditional FOG, the Shupe error of the same fiber length is about 1/6.5, and the Verdet constant of the fiber which is proportional to the Faraday error is 1/20 of the ordinary fiber. Compared with a traditional gyroscope, air-core photonic crystal fiber FOGs have great advantages in environmental adaptability and long-term stability.

In 2007, S. Blin of Ginzton Laboratory demonstrated that using an air-core fiber instead of a solid-core fiber dramatically reduces the thermal sensitivity of a FOG [5], as a result of both the reductions in Shupe constant (relative change of propagation phase with temperature) and the lower mode effective index. The diagram is shown in Figure 3.4. Experiments show that the thermal sensitivity of air-core fiber is lower than that of solid-core fiber by a factor of 6.5. They claimed that this property constitutes a significant improvement for the FOG.

3.2. Resonant Fiber-Optic Gyroscope

The resonant fiber-optic gyroscope (R-FOG) was discovered by professor S. Ezekiel of Massachusetts Institute of Technology (MIT). Compared with the laser gyroscope, the pump-less resonant cavity used by R-FOG can effectively avoid the optical locking effect. Compared with I-FOG, R-FOG requires shorter fiber length to achieve the same detection accuracy, and the thermal non-reciprocity is greatly reduced. At present, the research of R-FOG is not very mature, and it is still a long way from being practical. It is in the stage of transition from laboratory research to practical application.

3.2.1. Principle of R-FOG

The structure of R-FOG is shown in Figure 3.5. According to the conclusions in section 2.1, the optical path of clockwise and counterclockwise are not equal, thus the resonant frequencies of the two directions are uneven.

For the R-FOG, according to the theory of the annular cavity, the resonant frequency of the two traveling waves propagating in opposite directions in the annular cavity must be satisfied:

$$\nu = qc/L \tag{3.5}$$

where $q = L/\lambda$ is the total number, which is referred to as the longitudinal mode number of traveling wave. The wavelength λ is a function of the traveling wave resonant cavity length, and only those wavelengths that satisfy $\lambda = L/q$, those whose length L is exactly times its total length, can become resonant. From the above equation, we can know that the frequency changes with the change of distance, and its change rate is:

$$\Delta\nu = -\nu\Delta L/L = -c\Delta L/\lambda L \tag{3.6}$$

According to the above two formulas, the frequency difference between the clockwise and counterclockwise traveling waves of the ring traveling wave oscillator is:

$$\Delta\nu = (4A/\lambda L)\,\Omega \tag{3.7}$$

where, λ is the wavelength of a light wave inside the luminal cavity, and $4A/\lambda L$ is the ratio factor. The frequency difference of two beams of light and rotation velocity has a linear relationship, so measuring the R-FOG frequency offset $\Delta\nu$, we can get angular velocity Ω conversion.

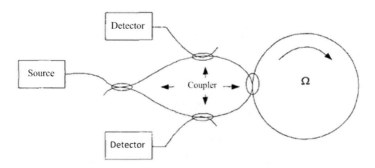

Figure 3.5. The principle diagram of the R-FOG.

3.2.2. Research Progress of R-FOG

Due to the above advantages, although I-FOG has become more and more mature, some large companies, research institutes and universities engaged in such research, have not abandoned R-FOG research, and made important progress in structural design, noise suppression and other technologies. Several foreign colleges engaged in the research of fiber-optic gyroscope, Massachusetts Institute of Technology, University of Tokyo, etc, have carried out the research of R-FOG.

In 1977, MIT's S. Ezekiel and S. K. Balsamo first proposed R-FOG. The advantage of R-FOG is that the length of the fiber is relatively short, usually only a few meters to a few tens of meters. However, it is still in the laboratory research stage due to the influence of Rayleigh scattering and the Kerr effect.

In 1982, L. F. Schalke et al. from Stanford University established the FRR model and developed the single-mode fiber ring resonator. In the same year, R. E. Meyer et al. [6] of MIT designed the first R-FOG, as shown in Figure 3.6 below. The FRR Finesse of the prototype reached 140. Under the integral condition of 1s, the zero-bias stability of 100s is $0.5(°)/h$.

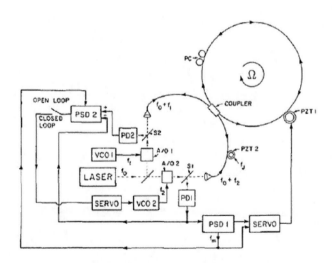

Figure 3.6. R-FOG based on single-mode fiber ring resonator proposed by L. F. Schalke et al.

In 1988, G. A. Sanders et al. [7] of the American company Honeywell, first used LiNbO₃ based Phase Modulation (PM) for light wave Modulation, and formed the R-FOG closed-loop technology scheme using bias maintaining FRR. Three years later, they used a single point of 90° polarization axis welding, polarization-maintaining, FRR constructed transmission type R-FOG, as shown in the Figure 3.7 below. It eliminates the FRR caused by temperature change in the polarization fluctuations, measured gyroscope zero bias stability reached the 6 $(°)/h$. [8]

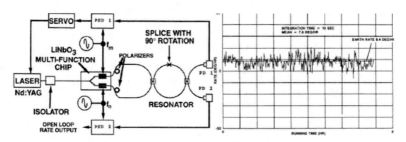

Figure 3.7. The diagram of weld polarization-maintaining FRR transmission type R - FOG block based on a single point of 90° polarization rotation axis and the test results.

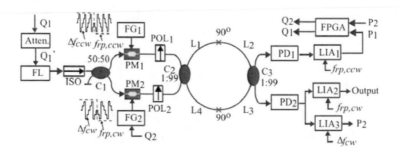

Figure 3.8. The diagram of the measurement based on double cavity 90° polarization axis welding FRR polarization fluctuation.

In 1997, Japanese scholar K. Hotate realized the fully digital closed-loop detecting R-FOG based on digital oblique wave phase modulation. It uses optical waveguide PM to replace the traditional light frequency shift device (Acousto Optic Frequency Shifter, AOFS), to improve the system integration and miniaturization. In combination with double frequency modulation for the first time, the backscattering effectively suppressed noise and improved the output SNR of the system. Then, in 2010, he proposed the intracavity double 90° polarization axisymmetric weld scheme, compared with the single point of 90° polarization axis fusion scheme, which more effectively suppressed the polarization fluctuation noise caused by drift, as shown in the Figure 3.8 above [9].

In the second generation of FOG, the development of R-FOG is more promising. In theory, it has been shown to be more accurate than I-FOG. Its general direction of development is the same as that of I-FOG, and it can be used not only in aerospace, but also in civil applications. As shown by the studies on R-FOG, the main technical problems are as follows:

1. Laboratory prototypes mostly adopt Nd:YAG solid state laser or gas laser and acousto-optic modulators (AOM) and other discrete devices, which cannot realize miniaturization and full packaging.
2. In principle, the higher the precision of the resonator, the more sensitive the FOG. Although fiber-optic resonator (F > 1000) with high precision has been reported, the precision of R-FOG sample in the laboratory is still very low (generally less than 100), which limits the performance of R-FOG.
3. Which modulation/demodulation scheme is more advantageous remains to be determined and further research is needed.
4. The suppression of temperature drift, back-Rayleigh scattering, polarization-related noise and optical Kerr effect and other noises requires different debugging or feedback circuits, which need to be optimized and further improved.

3.3. Stimulated Brillouin Scattering Fiber-Optic Gyroscope

The angular measuring device that is based on the frequency change principle of Brillouin fiber ring resonator is called Stimulated Brillouin scattering fiber-optic gyroscope (B-FOG).

3.3.1. Principle of B-FOG

The structure of B-FOG is similar to that of R-FOG. The basic principle of B-FOG is shown in Figure 3.9. The pumped light emitted by high power narrow-spectrum laser is divided into two equal-intensity beams after passing through coupler 1. After passing through coupler 2, most of the light enters the optical fiber ring. When the power reaches the threshold power of stimulated Brillouin scattering, the pumped light will excite the opposite stimulated Brillouin scattering light (Stokes light). Because of the influence of the Sagnac effect, the frequency of scattered light varies with the rotation angular velocity of the fiber ring.

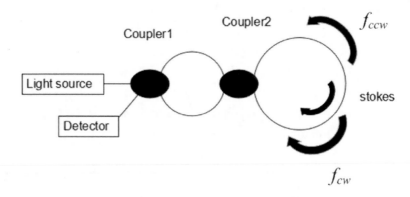

Figure 3.9. The principle diagram of the B-FOG.

The frequency f_{ccw} of scattered light wave along the counter-clockwise (CCW) direction and the frequency f_{cw} of scattered light wave along the clockwise (CW) direction change equally, but the symbol is opposite. The beat frequency difference between the two is proportional to the frequency $(f_{cw} - f_{ccw})$ of the detected signal and the rotation angular velocity of the fiber ring. Therefore, the rotation angular velocity of the optical fiber ring can be obtained by detecting the beat frequency signal and extracting the frequency. The relationship between the rotation angular velocity of the gyro and the beat frequency of two Stokes beams is deduced as follows:

$$\Omega = \frac{\lambda n L}{4A} \Delta v$$

(3.8)

where, L is the length of fiber coil, n is the refractive index of light, λ is the wavelength of incident pump light, A is the area of fiber coil.

3.3.2. Research Progress of B-FOG

At present, many people have studied B-FOG, such as Stanford University and Massachusetts Institute of Technology, mainly on the polarization, locking, orientation, Kerr effect and other noise sources of Brillouin FOG. In China, the threshold optical power and polarization of B-FOG have been theoretically analyzed by Northern Jiaotong University, and good research results have been achieved. However, as the third generation of FOG, the theory of B-FOG is not perfect, so most people are still in the research stage.

In 1980, P. J. Thomas, Van Driel and Stegmans proposed that a Brillouin fiber-optic gyroscope could be used to measure the angular velocity of inertia, marking the birth of B-FOG. At that time, there were some limitations in the experiment. They did not observe the beat effect of two Brillouin scattered light beams at a certain angular velocity.

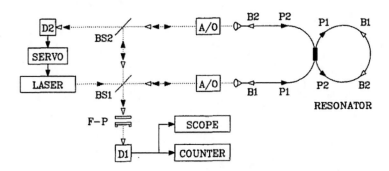

Figure 3.10. The BFOG experimental setup of F. Zarinetchi at the Massachusetts Institute of Technology.

In 1991, the MIT, F. Zarinetchi team measured the beat frequency of Brillouin scattered light through experiments[10]. The schematic diagram of the experimental device of the research group is shown in Figure 3.10. The actual maximum beat frequency of two SBS light is 3 kHz, which is close to the theoretical beat frequency of 3.2 kHz. However, when the rotating speed of the turntable is lower than a certain value, the output beat frequency is zero, and the locking phenomenon of the gyroscopy is detected in the experimental test. This is due to the coupling of two SBS beams with similar frequencies due to the backscattering of two SBS beams in the cavity at low rotational speed, which leads to frequency traction and makes the two SBS beams locked at the same frequency.

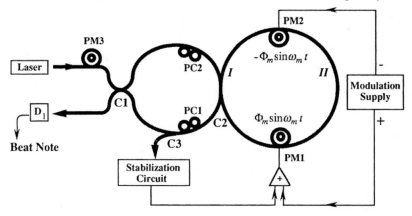

Figure 3.11 The BFOG system of Stanford University.

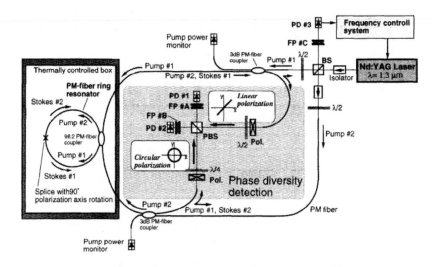

Figure 3.12. The scheme of BFOG with directional sensitivity of Tokyo University.

In 1993, the H. J. Shaw team of Stanford University made a systematic study of BFOG and proposed an optical dithering device to eliminate the lock-in effect of gyroscope at low speed [11]. As shown in Figure 3.11, in the system scheme, two piezoelectric ceramic phase modulators PM1 and PM2 are used to modulate the light transmitted in both directions. The length of the fiber is wound on each piezoelectric ceramic column. The two piezoelectric ceramic phase modulators are in push-pull working state under the control of the modulation signal. The optical path "Jitter" is realized to achieve the effect of mechanical jitter in a laser gyroscope.

In 1996, the K. Hotate group of Tokyo University reported their B-FOG device with direction discrimination function [12]. The schematic diagram of the scheme is shown in Figure 3.12. In the scheme, the optical fiber ring cavity is placed in the thermostat to avoid the influence of temperature fluctuation on B-FOG, and the feedback loop is added to the pump light source for frequency stabilization control. When the rotation angular velocity of the gyroscope is +0.18°/s, the detected beat frequency is 440 Hz, and 320Hz while the rotation angular velocity is -0.13°/s.

3.4. Slow Light Gyroscope

With the continuous development of optoelectronic technology, people have discovered the existence of the "slow light" effect, and also found the potential value of the slow light effect in optical communication technology, so people began to be interested in the study of slow light. In 1990, Harris et al. [13] proposed that electromagnetic transparency could be used to slow down the speed of light and hardly be absorbed. In 1999, Hau et al. [14] reduced the group velocity of light to 17 m/s by propagating light in ultra-cold atomic gas, which was a very significant breakthrough. In 2003, Bajcsy et al. [15] controlled the speed of light within a few hundred milliseconds by using the method of generating atomic spin coherence using light pulses with fixed envelope boundaries. In 2003, Bigelow et al. [16] first used the coherent population oscillation (CPO) technique to slow the speed of light to 57.5 m/s in ruby crystals at room temperature. In 2005, Longdell et al. [17] used the principle of electromagnetic induction induced transparency to doped rare earth elements in silicates, and realized the "freezing" phenomenon of maintaining the storage time of light for one second.

3.4.1. Debate on the Mechanism of Sensitivity Enhancement of Slow Light Gyroscope

In order to solve the "bottleneck" of the sensitivity of the gyroscope, people try to apply the slow light effect to the gyroscope, so as to improve the performance of the gyroscope. An important conclusion is drawn from the derivation of the last chapter: The phase shift does not depend on the shape of the area, the location of the center of rotation, and the presence of the refracting medium in the waveguide.

There had been a historical dispute on the interplay between the Sagnac effect and the Fresnel drag effect since the first demonstration of an operating fiber-optic gyroscope was reported by Vali and Shorthill. Leonhardt and Piwnicki [18] first theoretically proved that the "slow light" property induced by electromagnetically induced transparency (EIT) and coherent population trapping (CPT) may greatly boost the gyroscope's sensitivity by as much as the light slows. Zimmer et al. [19] then refuted the results of Leonhard and Piwnicki, and demonstrated that Sagnac phase shifts in dispersive media (the extra phase shifts generated by electromagnetic waves traveling in moving media versus those traveling in stationary media) are independent of group velocity. Then, Shahriar [20] proved that slow light in dispersion medium could enhance the sensitivity of relative rotation (the medium moved relative to the rotating reference system) by using doppler frequency shift, but could not enhance the sensitivity of absolute rotation (the medium was relatively static relative to the rotating reference system). Peng et al. [21] pointed out that the dispersive medium such as EIT cannot be utilized to enhance the absolute rotation-induced Sagnac phase shift, but the dispersive structure with slow light property is still beneficial to the enhancement of the Sagnac effect, since the structure's response is susceptible to the phase perturbation that was induced by the pure Sagnac effect.

3.4.2. Sagnac Effect in a Highly Dispersive Resonator Structure

Since the the dispersive medium cannot be used to enhance the absolute rotation-induced Sagnac phase shift, it cannot be utilized for navigation purposes. However, a dispersive structure can make up for this shortcoming. In this section, we will discuss the principle of the slow light gyroscope based on dispersive structure.

We use the theoretical model provided in [21] to elucidate the mechanism of the slow light gyroscope.

Assume that there is a single-ring resonator whose radius is R, and the refractive index is n. When the beam couples into the resonator, it resonates in the cavity. The structure makes the beam circulate in the closed loop cavity time after time, and every circulation contributes a phase shift because of the Sagnac effect. Since the resonator structure is highly dispersive, the response is susceptible to the phase shift. Therefore, we expect that the resonator structure will achieve a highly sensitive rotation sensor with compact size.

The response of the resonator can be described by the transfer function as:

$$H(\omega) = A(\omega)\exp[j\Phi(\omega)]$$

$$(3.9)$$

where A(ω) is the response of the amplitude, and $\Phi(\omega)$ is the response of the phase. Assuming that the single-ring resonator is lossless, there is no amplitude difference between input light and output light, hence A(ω) $\equiv$ 1. Then, the phase response can be expressed as:

$$\Phi(\omega) = \tan^{-1}\{\frac{\text{Im}[H(\omega)]}{\text{Re}[H(\omega)]}\}$$

$$(3.10)$$

The resonator is always structured by a series of basic elements, and then H(ω) can be written in the form of every basic element's transfer function. We are interested in a phase response rather than a whole transfer function, and we then rewrite Eq. (3-10) as:

$$\Phi(\phi(\omega)) = \tan^{-1}\{\frac{\text{Im}[H(\phi(\omega))]}{\text{Re}[H(\phi(\omega))]}\} \tag{3.11}$$

where $\phi(\omega)$ is the phase response for a single element. In any medium, the absolute rotation related phase shift of single segment $d\vec{r}$ is given by:

$$d(\Delta\phi) = \frac{\omega n^2}{c^2}(1-\alpha)\vec{V}\cdot d\vec{r} \tag{3.12}$$

where α is the Fresnel-Fizeau drag coefficient given by $\alpha = 1 - n^{-2}$, and $\vec{V}$ is the linear velocity of the segment. The Sagnac phase shift for a element can be calculated from Eq. (3-12). Unlike the dispersive medium, the phase shift in dispersive structure is induced by the pure Sagnac effect of an absolute rotation, and it is in no way related to both the material and the waveguide dispersion of the light path.

The phase shift of two counterdirection beams in the resonator can be calculated as:

$$\Delta\Phi = \Phi_+[(\phi(\omega))+\Delta\phi] - \Phi_-(\phi(\omega)) = \frac{\partial\Phi/\partial\omega}{\partial\phi/\partial\omega}\Delta\phi \tag{3.13}$$

where $\Delta\phi$ is the phase difference of the counterdirections in a single element induced by the Sagnac effect. The total phase shift $\phi(\omega)$ contains two parts, which are the phase shift from propagating and the Sagnac effect. The total phase shift of CW beam $\phi^+(\omega)$ and CCW beam $\phi^-(\omega)$ are:

$$\phi^+(\omega) = \frac{n\omega L}{c} + \frac{2\omega A}{c^2}\Omega \tag{3.14}$$

$$\phi^-(\omega) = \frac{n\omega L}{c} - \frac{2\omega A}{c^2}\Omega \tag{3.15}$$

where L and A are the length and the area of the resonator, respectively. For small Ω, $\Omega R \ll c$, neglecting the second terms, then Eq. (3-13) can be rewritten as:

$$\Delta\Phi(\omega) = \frac{\partial\Phi/\partial\omega}{nL/c}\Delta\phi = \frac{c/n}{L/(\partial\Phi/\partial\omega)}\Delta\phi \qquad (3.16)$$

As $\tau(\omega) = -(\partial\Phi/\partial\omega)$ is the group delay of the resonator, $v_g = L/\tau(\omega)$ is the group velocity and $n_g = c/v_g$ is the group index of the system, we get the phase shift for $\Delta\phi = (4\omega A/c^2)\Omega$:

$$|\Delta\Phi(\omega)| = \frac{4\omega A}{c^2}\Omega\cdot\frac{c}{v_g}\frac{1}{n} = \frac{4\omega A}{c^2}\Omega\cdot\frac{n_g}{n} \qquad (3.17)$$

Eq. (3-17) shows that the sensitivity of the rotating sensor is proportional to n_g; here ng represents the dispersion property that is contributed by the whole system. n_g is a key property for evaluating the performance of the resonator and the optimizing aim to design a highly dispersive resonator structure, although it is not the essential cause of the increase in sensitivity.

3.4.3. Research Progress of a Slow Light Gyroscope

Since the dispersion structure can improve the absolute rotation sensitivity, it can be applied to navigation. As a typical dispersion structure, coupled resonators have attracted more and more attention due to their unique slow light characteristics.

In 2004, Matsko et al. [22] first proposed the use of coupled resonator to improve the sensitivity of the rotation sensor. Figure 3.13 shows the coupled resonator rotation sensor designed by Matsko. In the sensor, the Sagnac effect, caused by rotation, is enhanced by the resonator. However, the problem with this type of sensor is that the Sagnac effect between the resonators is cancelled out because the light waves travel in opposite directions in the internal and external resonators of the annular waveguide, reducing the sensor's sensitivity (the transmission direction of the resonator inside the annular waveguide is clockwise, while the transmission direction of the resonator outside the annular waveguide is counterclockwise). Later, Matsko [23] realized the above problem and improved the structure, pointing out that the resonator was either inside or outside the annular waveguide. Although there are some deficiencies in Matsko's structural design and theoretical analysis, it provides other researchers with the idea of using a coupled resonator to improve the sensitivity of the rotation sensor.

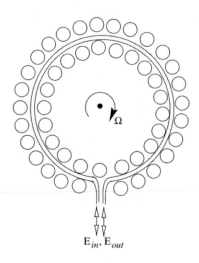

Figure 3.13. The rotation sensor made of coupled resonators proposed by Matsko.

Figure 3.14 illustrates the coupled resonator rotation sensor proposed by Scheuer and Yaris [24]. The light wave is input by the input waveguide and evenly divided into two beams transmitted in the opposite direction by a 3dB coupler. The two beams are transmitted through the coupled resonator, and finally converge and interfere at the 3dB coupler. The interference signal is output by the two ports of the 3dB coupler. The light intensity after interference between the two ports of the coupler depends on the phase difference of the two beams. The phase difference between the two beams is caused by the Sagnac effect in the coupled resonator, so the rotation velocity can be obtained by measuring the variation of the interference light intensity at either port. According to the numerical results of Scheuer and Yariv, the sensitivity of the coupled resonator rotation sensor is much higher than that of the conventional rotation sensor of the same length. Meanwhile, they pointed out that for practical applications, the coupled resonator rotation sensor has the following advantages: (1) The output signal depends on the coupling between the resonators, which makes the device rotation sensing sensitivity independent of the device area. (2) The responsiveness does not depend on the resonator layout, so the chip utilization efficiency is greatly increased.

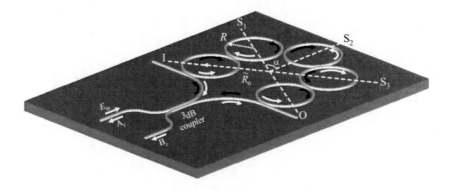

Figure 3.14. The rotation sensor made of coupled resonators proposed by Scheuer and Yaris.

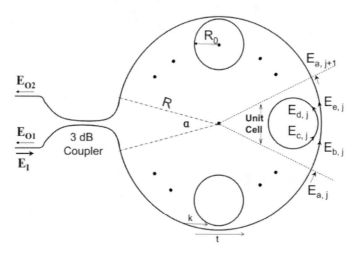

Figure 3.15. The rotation sensor made of coupled resonators proposed by Yan.

Figure 3.15 shows the coupled resonator rotation sensor proposed by Yan [25]. By measuring the interference intensity of the two phase transmitted light waves in the coupled resonator, the rotation velocity is obtained. Theoretically, the differential form of phase difference of the two backpropagation waves at the resonance frequency of the coupled resonator is obtained. The result also shows that the sensitivity of the rotation sensor can be improved compared with the traditional one.

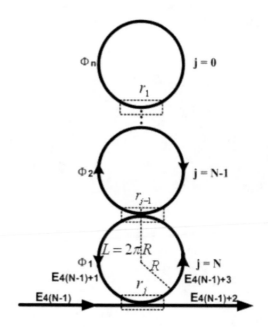

Figure 3.16. The rotation sensor made of coupled resonators proposed by Peng.

Then, Peng [21] put forward in theory that the gyroscope with high integration and high sensitivity can also be made with a series coupling ring resonator (as shown in Figure 3.16). Peng established a theoretical model describing the rotation-sensing characteristics of the gyroscope based on a series-coupled resonator structure, revealing that the Sagnac phase shift induced by rotation in such a ring resonator structure is proportional to the group refractive index of the resonator structure. With the same fiber length, the Sagnac phase shift can be obtained using the resonator structure, which is hundreds of times that obtained by the traditional I-FOG.

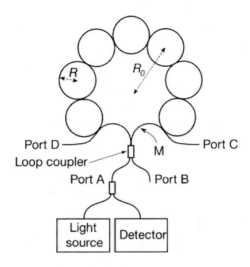

Figure 3.17. The rotation sensor made of coupled resonators proposed by Terrel.

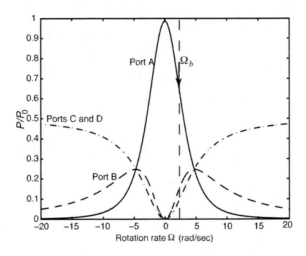

Figure 3.18. Response of the series-coupled ring resonator based gyroscope to rotation rate.

Then, Terrel [26] calculated the relationship between the interference light intensity and the rotation velocity of the gyroscope (structure as shown in Figure 3.17) based on the series coupled resonator. As shown in

Figure 3.18, the sensitivity of the sensor is dependent on the fineness of the resonator.

In order to improve the sensitivity of the gyroscope based on the resonator, Sorrentino [27] proposed using the serially coupled ring resonator with periodic coupling coefficient to conduct rotation sensing. The basic structure is shown in Figure 3.19. The theoretical calculation shows that the inertial-level rotation velocity ($0.001°/h$) can be sensed on an area less than $0.1mm^2$ by using this structure. It is shown that the series coupled ring resonator can be used to realize a gyroscope with high integration and sensitivity.

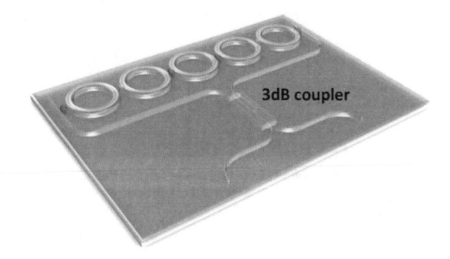

Figure 3.19. The rotation sensor made of coupled resonators proposed by Sorrentino.

From the above studies on the application of slow light in the coupled resonator waveguide in the aspect of rotation sensing, the research results are mostly theoretical, with few experimental reports. Meanwhile, the number of coupled resonators in the theoretical model is too large, which greatly increases the difficulty of experimental verification. Moreover, from the perspective of practical application, more resonators will increase the complexity of the structure of the rotating sensor, bringing many unstable factors to the system, which is not conducive to practical application.

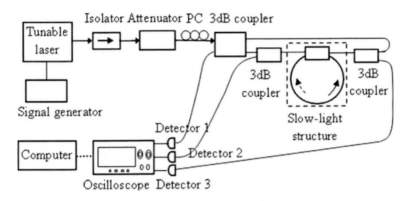

Figure 3.20. Experiment setup for the measurement of the slow light gyroscope by Y. Zhang.

In terms of experiments, Y. Zhang et al. [28] from Harbin Institute of Technology carried out experimental verification of the ring resonator-based I-FOG (as shown in Figure 3.20), which proved that the high dispersion (high group refractive index) of the phase response of the ring resonator could be used to realize the I-FOG with high sensitivity and high integration.

Then, H. Tian [29] further confirmed that the I-FOG with high sensitivity and integration can also be realized by using the high dispersion of the parallel coupled ring resonator structure through experiments (as shown in Figure 3.21). As shown in Figure 3.22, the light intensity at the resonant frequency increases with the rotation rate.

In 2011, X Zhang [30] broke through the limit of sensitivity-line width product by using ring resonators with dynamically adjustable bandwidth, and realized highly sensitive rotation sensing without using a laser light source with narrow line width.

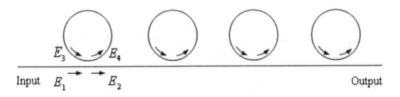

Figure 3.21. The rotation sensor made of coupled resonators proposed by H Tian.

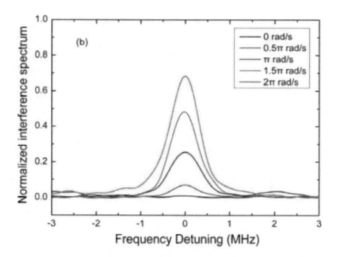

Figure 3.22. Experimental interference spectra of one resonator at different rotary velocities: 0 rad/s, 0.5π rad/s, 1π rad/s, 1.5π rad/s, and 2π rad/s.

In 2018, H. Tian [31] proposed a rotation sensing frame based on the self-interference add–drop resonator (as shown in Figure 3.23), in which two waves spontaneously form and their interference spectra contain angular velocity information. Moreover, the direction of angular velocity can be recognized by setting the asymmetric coupling coefficient in the resonator.

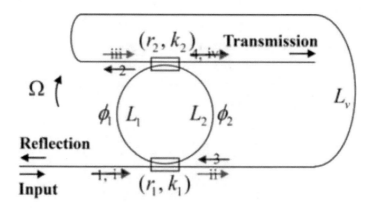

Figure 3.23. The rotation sensor based on self-interference add–drop resonator proposed by H Tian.

3.5. Conclusion

The research of I-FOG is very mature and has already appeared in commercial products, widely used in a variety of fields. At present, the research of I-FOG is mainly focused on miniaturization, integration and precision. For higher precision, photonic crystal fibers are introduced as waveguide coils for the I-FOGs. To achieve the same sensitivity, R-FOG needs shorter fiber length by the circular cavity than I-FOG. However, the research of R-FOG is still in the laboratory stage, and has not reached the stage of practical application. B-FOG greatly improves the sensitivity of the gyroscope based on the stimulated Brillouin scattering principle. Compared with R-FOG, the structure of B-FOG components are reduced by half. Meanwhile, B-FOG can directly produce a gyroscope beat frequency, without the complex modulation and demodulation of R-FOG, thus it shows more practical prospects.

In essence, it is not the "light slows" that boost the slow-light gyroscope's sensitivity. So, strictly speaking, the name "slow-light gyroscope" is not very accurate. Nevertheless, group velocity reduction is indeed an important phenomenon in positive dispersion, and the sensitivity of the rotating sensor is proportional to the group velocity. Therefore, the academic circle still adopts the term "slow-light gyroscope." The dispersive medium could enhance the sensitivity of relative rotation, but cannot be utilized to enhance the absolute rotation-induced Sagnac phase shift. The dispersive structure can enhance the sensitivity of absolute rotation, so it can be applied to navigation. As a typical dispersion structure, coupled resonators have the advantage of lower cost and easier integration, and the possibility of improving the sensitivity of a slow light gyroscope has been demonstrated experimentally for the first time by researchers from Harbin Institute of Technology.

4. FUNCTIONS AND APPLICATIONS

Gyroscopes are widely used in many fields. As signal sensors, they can provide information such as attitude, speed and position, and provide navigation or guidance for navigation bodies and weapon systems. As a stabilizer, the gyroscope can be mounted on a variety of carriers to help reduce swaying and achieve stability. As a precision test instrument, the gyroscope can provide accurate azimuth datum for ground facilities, mining tunnels, underground railways, oil drilling and missile silos.

The inertial navigation system (INS) with a gyroscope and accelerometer as main inertial components is a typical application of a gyroscope. INS is an autonomous navigation device that can provide continuous and real-time information of a carrier's position, attitude and speed. Its main feature is that it does not rely on external information, climate conditions and external factors to interfere. At the same time, it does not radiate energy to the outside, has a good concealment, and makes up for the global satellite navigation system relying on radio waves, poor anti-interference ability, anti-deception ability. Therefore, the use of inertial/satellite integrated navigation technology can achieve high-precision continuous navigation worldwide, and the inertial navigation system output signals have continuity and universality, so that the carrier can work continuously at any time and any place. The integrated navigation based on inertial technology is the research focus of navigation technology and will be further developed towards multi-sensor fusion.

Inertial navigation and control systems are the core technology product of modern national defense systems. With the decreased cost and the increased demand, the application field of inertial navigation technology is broadened and plays an increasingly important role in the national economy. It has been applied in commercial fields such as civil carrier vehicle, resource exploration, ocean exploration, geodetic survey and tunnel railway construction. In the future, it will serve people's production and life more conveniently and effectively in the fields of consumer electronics, medical electronics and indoor navigation.

ACKNOWLEDGMENTS

The project is supported by Shanghai Aerospace Science and Technology Innovation Fund, grant no. SAST2017-099 and the National Key R&D Program of China (no. 2018YFC1503703).

REFERENCES

[1] Vali V., Shorthill R. W. 1976. "Fiber ring interferometer." *Applied Optics* 15:1099-1100. doi: 10.1364/AO.15.001099.

[2] Hervé C. Lefèvre. 1997. "Fundamentals of the interferometric fiber-optic gyroscope." *Optical Review* 4: A20-A27. doi: 10.1007/bf02935984.

[3] Jesse Tawney, Farhad Hakimi, R. L. Willig. 2006. "Photonic Crystal Fiber IFOGs." *OFS 19th International Conference on Optical Fiber Sensors* p.ME8. doi: 10.1364/OFS.2006.ME8.

[4] Kim H. K., Digonnet M. J. F., Kino G. S. 2006. "Air-core photonic-bandgap fiber-optic gyroscope." *Journal of Lightwave Technology* 24:3169-3174. doi: 10.1109/JLT.2006.880689.

[5] Kim H. K., Dangui V., Digonnet M., et al. 2007. "Reduced Thermal Sensitivity of a Fiber-Optic Gyroscope Using an Air-Core Photonic-Bandgap Fiber." *Journal of Lightwave Technology* 25: 861-865. doi: 10.1109/JLT.2006.889658.

[6] Meyer R. E., Ezekiel S., et al. 1983. "Passive fiber-optic ring resonator for rotation sensing." *Optics Letters* 8:644-646. doi: 10.1364/OL.8.000644.

[7] Sanders G. A., Rouse G. F., Strandjord L. K., et al. 1989. "Resonator Fiber-Optic Gyro Using LiNbO3 Integrated Optics At 1.5-μm Wavelength." *Fiber Optic and Laser Sensors* VI 985:203-210. doi: 10.1117/12.948857.

[8] Strandjord L. K., Sanders G. A. 1992. "Resonator fiber optic gyro employing a polarization-rotating resonator." *Proceedings of SPIE -*

The International Society for Optical Engineering 1585:163-172. doi: 10.1117/12.135044.

[9]   Wang X., He Z., Hotate K. 2010 "Reduction of polarization-fluctuation induced drift in resonator fiber optic gyro by a resonator with twin 90 degrees polarization-axis rotated splices." *Optics Express* 18:1677-1683. doi: 10.1364/OE.18.001677.

[10]   Zarinetchi F., Smith S. P., Ezekiel S. 1991 "Stimulated Brillouin fiber-optic laser gyroscope." *Optics Letters* 16:229-231. doi: 10.1364/OL.16.000229.

[11]   Huang S., Toyama K., Nicati P. A., et al. 1992 "Brillouin fiber optic gyro with push-pull phase modulator and synthetic heterodyne detection." *Fiber Optic and Laser Sensors X* 1795:48-59. doi: 10.1117/12.141277.

[12]   Tanaka Y. T. Y., Yamasaki S. Y. S., Hotate K. H. K. 2002 "Brillouin fiber-optic gyro with directional sensitivity." *IEEE Photonics Technology Letters* 8:1367-1369. doi: 10.1109/68.536657.

[13]   Harris S. E., Field J. E., Imamoglu A. 1990 "Nonlinear optical processes using electromagnetically induced transparency." *Physical Review Letters* 64:1107-1110. doi: 10.1109/NLO.1990.695974.

[14]   Hau L. V., Harris S. E., Dutton Z., et al. 1999 "Light speed reduction to 17 metres per second in an ultracold atomic gas." *Nature* 397: 594-598. doi: 10.1038/17561.

[15]   Bajcsy M., Zibrov A. S., Lukin M. D. 2003 "Stationary pulses of light in an atomic medium." *Nature* 426: 638-641. doi: 10.1038/nature02176.

[16]   Boyd R. W., Bigelow M. S., Lepeshkin N. N. 2003 "Superluminal and ultra-slow light propagation in room-temperature solids." *Science* 301: 200-202. doi: 10.1126/science.1084429.

[17]   Longdell J. J., Fraval E., Sellars M. J., et al. 2005 "Stopped Light with Storage Times Greater than One Second Using Electromagnetically Induced Transparency in a Solid." *Physical Review Letters* 95: 063601. doi: 10.1103/PhysRevLett.95.063601.

[18] Leonhardt U., Piwnicki P. 2000 "Ultrahigh sensitivity of slow-light gyroscope." *Physical Review* A 62: 055801. doi: 10.1103/PhysRevA. 62.055801.

[19] Zimmer F., Fleischhauer M., Lukin M. D. 2004 "Sagnac interferometry based on ultra-slow polaritons in cold atomic vapors." *Physical Review Letters* 92:253201. 10.1103/Phys Rev Lett. 92. 253201.

[20] Shahriar M. S., Pati G. S., Tripathi R., et al. 2005 "Ultrahigh Precision Absolute and Relative Rotation Sensing using Fast and Slow Light." *Physical Review A*. doi: 10.1103/PhysRevA.75.053807.

[21] Peng C., Li Z., Xu A. 2007 "Rotation sensing based on a slow-light resonating structure with high group dispersion." *Applied Optics* 46:4125-4131. doi: 10.1364/AO.46.004125.

[22] A. B. Matsko, A. A. Savchenkov, V. S. Ilchenko, L. Maleki. 2004. "Optical gyroscope with whispering gallery mode optical cavities." *Optics Communications* 233:107-112. doi: 10.1016/j.optcom.2004. 01.035.

[23] Matsko A. B., Savchenkov A. A., Ilchenko V. S., et al. 2006. "Erratum to 'Optical gyroscope with whispering gallery mode optical cavities'." *Optics Communications* 259: 393-394. doi: 10.1016/j. optcom.2005.08.017.

[24] Shahriar M. S., Pati G. S., Tripathi R., et al. 2005. "Ultrahigh Precision Absolute and Relative Rotation Sensing using Fast and Slow Light." *Physical Review A*. doi: 10.1103/PhysRevA.75.053807.

[25] Yan L., Xiao Z., Guo X., et al. 2009. "Circle-coupled resonator waveguide with enhanced Sagnac phase-sensitivity for rotation sensing." *Applied Physics Letters* 95: 141104-1 - 141104-3. doi: 10.1063/1.3243456.

[26] Terrel M. A., Digonnet M. J. F., Fan S. 2009. "Performance Limitation of a Coupled Resonant Optical Waveguide Gyroscope." *Journal of Lightwave Technology* 27:47-54. doi: 10.1109/ JLT.2008.927753.

[27] Sorrentino C., Toland J. R. E., Search C. P. 2012. "Ultra-sensitive chip scale Sagnac gyroscope based on periodically modulated

coupling of a coupled resonator optical waveguide." *Optics Express* 20:354-363. doi: 10.1364/OE.20.000354.

[28] Yundong Zhang, He Tian, Xuenan Zhang, Nan Wang, Jing Zhang, Hao Wu, and Ping Yuan. 2010. "Experimental evidence of enhanced rotation sensing in a slow-light structure." *Optics Letters* 35:691-693. doi: 10.1364/OL.35.000691.

[29] Tian H., Zhang Y., Zhang X., et al. 2011. "Rotation sensing based on a side-coupled spaced sequence of resonators." *Optics Express* 19:9185-9191. doi: 10.1364/OE.19.009185.

[30] Zhang X., Zhang Y., Tian H., et al. 2011. "Ultrahigh sensitivity of rotation sensing beyond the trade-off between sensitivity and linewidth by the storage of light in a dynamic slow-light resonator." *Physical Review A* 84: 063823-1-063823-11. doi: 10.1103/PhysRevA.84.063823.

[31] Tian H., Zhang Y. 2018. "Rotation Sensing Based on the Sagnac Effect in the Self-Interference Add–Drop Resonator." *Journal of Lightwave Technology* 36:1792-1797. doi: 10.1109/jlt.2017.2788499.

In: Gyroscopes
Editor: Marcel Gerste

ISBN: 978-1-53615-856-4
© 2019 Nova Science Publishers, Inc.

Chapter 2

SIGNAL PROCESSING FOR MICROELECTROMECHANICAL SYSTEMS (MEMS) GYROSCOPES

Allen R. Stubberud[1] and Peter A. Stubberud[2]
[1]Department of Electrical Engineering and Computer Science,
University of California, Irvine, CA, US
[2]Department of Electrical and Computer Engineering,
University of Nevada, Las Vegas, US

ABSTRACT

For many systems requiring angular velocities of a distributed mass, such as guidance and control systems for unmanned space vehicles, it is highly desirable to have inertial measurement sensors that are small, inexpensive, low power, reliable and accurate. Technological advances in the design and construction of micro inertial sensors, such as accelerometers and gyroscopes, have much promise in providing small, inexpensive, and low power devices; however, additional improvement in the reliability and, especially, the accuracy of these micro devices is still necessary. Although major improvements in these two properties may occur in the future, in this chapter it is proposed that signal processing methods be used to provide appropriate accuracies and, in many cases, improved reliability. Specifically, it is proposed that appropriate gyroscope accuracy can be attained by using statistical methods to combine the output measurement of many, perhaps one hundred or more, MEMS gyrocopes on a single chip (or a few chips) to provide a single accurate measurement. One method of performing such a combination is through an extended Kalman filter (EKF). A standard application of an EKF to an array of gyroscopes would involve at least six state equations per gyroscope and the number of covariance equations would be in the order of the square of the product of six times the number of gyroscopes. Obviously, the 'curse of dimensionality' produces an explosion of computations. Even if the EKF for each individual gyroscope is uncoupled from the rest, the number of covariance equations is of the order of the number of gyroscopes times six squared, which can still lead to a formidable computational burden. In this chapter, two signal processing techniques are discussed, each for a different type of gyroscope noise. These two techniques are then combined to produce a general technique for improving the accuracy of a gyroscope. The final gyroscope output is generated by an EKF with a total of six state equations and (no more than) thirty six covariance equations.

1. INTRODUCTION

Navigation, guidance, and control systems for small space vehicles, such as satellites, require compact inertial measurement sensors to provide accurate position, velocity, and angular information about the vehicle. Recent developments in micro-electromechanical systems (MEMS) technology promise sufficiently small inertial sensors, but these sensors are not sufficiently accurate for many space applications. On the other hand, the cost of these sensors promises to be quite inexpensive; therefore, one method of improving sensor accuracy is by using many sensors to make measurements of the same quantity and then combining these measurements to generate one accurate measurement. Note that the use of many inexpensive micro-sensors measuring the same quantity can also provide reliability, through redundancy, at a reasonable cost. This latter point, while of considerable interest, is not directly addressed in this paper. The specific problem addressed here is that of combining the outputs of many micro-sensors, all measuring the same quantity, so that the accuracy of the combination greatly exceeds the accuracy of the individual micro-sensors but without requiring an extraordinary amount of computation. This is what is referred to in this chapter as the signal processing problem for MEMS gyroscopes. In this chapter, two signal processing techniques are discussed, each for a different type of gyroscope noise. In the first technique, the gyroscope output is only corrupted by white noise. In the second technique the gyroscope output is corrupted by random bias and manufacturing errors. These two techniques are then combined to produce a general technique for achieving improved gyroscope accuracy. The final gyroscope output is generated by an EKF with a total of six state equations and (no more than) thirty six covariance equations.

2. A MATHEMATICAL MODEL FOR A NOMINAL MEMS GYROSCOPE

Assume that a generic MEMS gyroscope design can be represented by a linear state vector form, that is, a set of n first-order linear differential equations (a set of state equations) given by;

$$\frac{d\mathbf{x}(t)}{dt} = \mathbf{A}\mathbf{x}(t) + \mathbf{b}z(t) \tag{1}$$

and a linear measurement equation:

$$y(t) = \mathbf{c}^T \mathbf{x}(t) \tag{2}$$

where $\mathbf{x}(t)$ is the n-dimensional system state vector, $z(t)$, is the scalar system input, $y(t)$ is the scalar system output, A is an $n \times n$ matrix of constant coefficients, $\mathbf{b}$ and $\mathbf{c}$ are $n \times 1$ vectors of constant coefficients. The system state vector $\mathbf{x}(t)$ and the system output $y(t)$ can also be written in terms of convolution integrals,

$$\mathbf{x}(t) = \int_{\tau=-\infty}^{t} \mathbf{H}(t-\tau)\mathbf{b}z(\tau)d\tau = \int_{\tau=-\infty}^{t} \mathbf{h}_\mathbf{b}(t-\tau)z(\tau)d\tau \qquad (3)$$

and

$$y(t) = \int_{\tau=-\infty}^{t} \mathbf{c}^T \mathbf{H}(t-\tau)\mathbf{b}z(\tau)d\tau = \int_{\tau=-\infty}^{t} \mathbf{h}_\mathbf{c}^T(t-\tau)\mathbf{b}z(\tau)d\tau \qquad (4)$$

where $\mathbf{H}(t-\tau) = \exp[\mathbf{A}(t-\tau)]$, $\mathbf{h}_\mathbf{b}(t-\tau) = \mathbf{H}(t-\tau)\mathbf{b}$, and $\mathbf{h}_\mathbf{c}(t-\tau) = \mathbf{H}^T(t-\tau)\mathbf{c}$. The gyroscope design values of the coefficients in these two equations will be called the *nominal coefficients.* A state equation and an output equation defined by the nominal coefficients will be called *nominal state and output equations* and together represent a *nominal MEMS gyroscope.* Manufactured versions of the designed gyroscope will deviate from the nominal system in that the actual coefficients of the manufactured *real gyroscopes* will deviate from the nominal coefficients. In addition, the outputs of the real gyroscopes will be corrupted by measurement noises, typically a white noise and a random bias in the output.

3. A SIGNAL PROCESSING TECHNIQUE FOR WHITE NOISE

As the first step in the general signal processing technique discussed in this chapter, consider a gyroscope, in which the gyroscope output $y(t)$ consists of the sum of the *true* output generated by the input $z(t)$ and an additive white noise. That is, consider a gyroscope defined by the state equation

$$\frac{d\mathbf{x}(t)}{dt} = \mathbf{A}\mathbf{x}(t) + \mathbf{b}z(t) \qquad (5)$$

and the output equation:

$$y(t) = \mathbf{c}^T \mathbf{x}(t) + v(t) \tag{6}$$

where $v(t)$ is a zero mean scalar white noise with covariance R. A standard Kalman filter could be applied to Equations (5) and (6) to generate an optimal estimate of the *true* state $\mathbf{x}(t)$ that is generated only by $z(t)$ and an optimal estimate of the *true* output (that part of $y(t)$ generated by the true state).

Now assume that a set of N gyroscopes has been chosen from an available sample space of gyroscopes. Each gyroscope is defined by the state equation

$$\frac{d\mathbf{x}_i(t)}{dt} = \mathbf{A}\mathbf{x}_i(t) + \mathbf{b}z(t) \tag{7}$$

Obviously, the states of all the gyroscopes are equal, that is, $\mathbf{x}_1(t) = \mathbf{x}_2(t) = \cdots = \mathbf{x}_N(t) = \mathbf{x}(t)$ where $\mathbf{x}(t)$ is the true state of the gyroscope. However, the states of these gyroscopes are assumed to be measured through a set of N noisy measurement equations

$$y_i(t) = \mathbf{c}^T \mathbf{x}_i(t) + v_i(t), \quad i = 1, 2, \cdots, N \tag{8}$$

where $v_i(t)$ is a zero mean scalar white noise associated with the ith measurement. The scalar white noises $v_i(t)$, $i = 1, 2, \cdots, N$ are assumed to be mutually independent and identically distributed each with covariance R. Note that the true state and the true output are common to each member of the set of gyroscopes. Taking the arithmetic average of the set of state equations (7) generates the following equation.

$$\frac{d}{dt}\left(\frac{1}{N}\sum_{i=1}^{N}\mathbf{x}_i(t) \right) = \mathbf{A}\left(\frac{1}{N}\sum_{i=1}^{N}\mathbf{x}_i(t) \right) + \mathbf{b}z(t) \tag{9}$$

which can be rewritten

$$\frac{d\mathbf{x}(t)}{dt} = \mathbf{A}\mathbf{x}(t) + \mathbf{b}z(t) \tag{10}$$

Similarly, an arithmetic average is taken of the set of measurement equations (8) generating the equation

$$\overline{y}(t) = \frac{1}{N}\left(\sum_{i=1}^{N} y_i(t)\right) = \mathbf{c}^T\left(\frac{1}{N}\sum_{i=1}^{N} \mathbf{x}_i(t)\right) + \frac{1}{N}\left(\sum_{i=1}^{N} v_i(t)\right) \tag{11}$$

which is rewritten

$$\overline{y}(t) = \mathbf{c}^T\mathbf{x}(t) + \overline{v}(t) \tag{12}$$

The elements of the summation $\overline{v}(t)$ are independent and identically distributed each with covariance R, therefore $\overline{v}(t)$ has a covariance matrix R/N. A standard Kalman filter can be applied to Equations (10) and (12) to generate an optimal estimate of the true state and the true output of the average gyroscope which are also the true state and true output of each gyroscope in the set of N gyroscopes. The state and output estimates for the average gyroscope are (probabilistically) better than the estimates for the individual gyroscopes because the covariance of the white noise is reduced by a factor of N in the average output. Apparently, if the other sources of error in the gyroscope output could be reduced to white noises, then this averaging technique could be used to reduce the output error. This idea is developed in the next sections of the chapter.

4. A REAL GYROSCOPE AND THE RANDOM PERTURBATION IN ITS OUTPUT

Consider a *real* gyroscope defined by the state equation (1) and output equation (2) but where the system coefficients are, due to manufacturing errors, perturbed from the nominal coefficients. It is assumed that no white noise and no random bias is present in the gyroscope output. The real gyroscope can be represented by a state equation given by:

$$\frac{d[\mathbf{x}(t) + \Delta\mathbf{x}(t)]}{dt} = (\mathbf{A} + \Delta\mathbf{A})[\mathbf{x}(t) + \Delta\mathbf{x}(t)] + (\mathbf{b} + \Delta\mathbf{b})z(t) \tag{13}$$

and an output equation given by:

$$(y(t) + \Delta y(t)) = (\mathbf{c} + \Delta\mathbf{c})^T (\mathbf{x}(t) + \Delta\mathbf{x}(t)) \tag{14}$$

where $\Delta\mathbf{x}(t)$ and $\Delta y(t)$ are the random variations in the state and the output generated by (matrix and vector) random variations in the coefficients, $\Delta\mathbf{A}$, $\Delta\mathbf{b}$, and $\Delta\mathbf{c}$. Assuming that the second order terms in Equations (13) and (14) are negligible compared to other terms in the equations, two linear equations defining the (vector and scalar) random variations, $\Delta\mathbf{x}(t)$ and $\Delta y(t)$, are given by:

$$\frac{d\Delta\mathbf{x}(t)}{dt} = \mathbf{A}\Delta\mathbf{x}(t) + \Delta\mathbf{A}\mathbf{x}(t) + \Delta\mathbf{b}z(t) \tag{15}$$

and

$$\Delta y(t) = \mathbf{c}^T \Delta\mathbf{x}(t) + \Delta\mathbf{c}^T \mathbf{x}(t) \tag{16}$$

respectively. Equation (16) can also be written as

$$\Delta y(t) = \Delta y_A(t) + \Delta y_b(t) + \Delta y_c(t) \tag{17}$$

where $\Delta y_A(t)$, $\Delta y_b(t)$, and $\Delta y_c(t)$ are the random perturbations in the gyroscope output due to the random perturbations ΔA, Δb, and Δc respectively. Using convolution integral forms and the definitions below Equation (4), the random perturbations in the output can be written as:

$$\Delta y_A(t) = \int_{\tau=-\infty}^{t} \left[\int_{\theta=\tau}^{t} \mathbf{h}_c^T(t-\theta)\Delta\mathbf{A}\mathbf{h}_b(\theta-\tau)d\theta \right] z(\tau)\,d\tau$$

$$\Delta y_b(t) = \int_{\tau=-\infty}^{t} \mathbf{h}_c^T(t-\tau)\Delta\mathbf{b}z(\tau)d\tau$$

$$\Delta y_c(t) = \int_{\tau=-\infty}^{t} \Delta\mathbf{c}^T\mathbf{h}_b(t-\tau)z(\tau)d\tau$$

Now let:

$$\mathbf{h}_b(\theta-\tau) = \begin{bmatrix} h_{b1}(\theta-\tau) & \cdots & h_{bn}(\theta-\tau) \end{bmatrix}^T$$

$$\mathbf{h}_c(t-\theta) = \begin{bmatrix} h_{c1}(t-\theta) & \cdots & h_{cn}(t-\theta) \end{bmatrix}^T$$

and

$$\Delta\mathbf{A} = \begin{bmatrix} \Delta a_{11} & \Delta a_{12} & \cdots & \Delta a_{1n} \\ \Delta a_{21} & \Delta a_{22} & \cdots & \Delta a_{2n} \\ \cdots & \cdots & \ddots & \cdots \\ \Delta a_{n1} & \Delta a_{n2} & \cdots & \Delta a_{nn} \end{bmatrix}$$

then $\Delta y_A(t)$ can be written as:

$$\Delta y_A(t) = \sum_{i=1}^{n}\sum_{j=1}^{n} \Delta a_{ij} \int_{\tau=-\infty}^{t} \left[\int_{\theta=\tau}^{t} h_{ci}(t-\theta)h_{bj}(\theta-\tau)d\theta \right] z(\tau)d\tau$$

$$= \sum_{i=1}^{n}\sum_{j=1}^{n} \Delta a_{ij} \int_{\tau=-\infty}^{t} h_{cbij}(t-\tau)z(\tau)d\tau$$

$$= \sum_{i=1}^{n}\sum_{j=1}^{n} \Delta a_{ij} \int_{\tau=0}^{\tau=\infty} h_{cbij}(\tau)z(t-\tau)d\tau$$

where

$$h_{cbij}(t - \tau) = \int_{\theta=\tau}^{t} h_{ci}(t - \theta)h_{bj}(\theta - \tau)d\theta.$$

Similarly, $\Delta y_b(t)$ and $\Delta y_c(t)$ can be written as:

$$\Delta y_b(t) = \sum_{i=1}^{n} \Delta b_i \int_{\tau=-\infty}^{t} h_{ci}(t - \tau)z(\tau)d\tau$$

$$= \sum_{i=1}^{n} \Delta b_i \int_{\tau=0}^{\tau=\infty} h_{ci}(\tau)z(t - \tau)d\tau$$

$$\Delta y_c(t) = \sum_{i=1}^{n} \Delta c_i \int_{\tau=-\infty}^{t} h_{bi}(t - \tau)z(\tau)d\tau$$

$$= \sum_{i=1}^{n} \Delta c_i \int_{\tau=0}^{\tau=\infty} h_{bi}(\tau)z(t - \tau)d\tau$$

Substituting these results into (17) gives the following form for $\Delta y(t)$:

$$\Delta y(t) = \sum_{i=1}^{n}\sum_{j=1}^{n} \Delta a_{ij} \int_{\tau=0}^{\tau=\infty} h_{cbij}(\tau)z(t - \tau)d\tau + \sum_{i=1}^{n} \Delta b_i \int_{\tau=0}^{\tau=\infty} h_{ci}(\tau)z(t - \tau)d\tau + \sum_{i=1}^{n} \Delta c_i \int_{\tau=0}^{\tau=\infty} h_{bi}(\tau)z(t - \tau)d\tau$$

$$(18)$$

It is assumed for this real gyroscope, that all the coefficients in ΔA, Δb, and Δc are mutually independent random variables, are independent of the input signal, $z(t)$, and have the first and second order statistics

$$E[\Delta a_{ij}] = 0, \ \text{var}[\Delta a_{ij}] = \sigma_{a_{ij}}^2, \ i, j = 1, 2, \cdots n$$

$$E[\Delta b_i] = 0, \ \text{var}[\Delta b_i] = \sigma_{b_i}^2, \ i = 1, 2, \cdots n$$

$$E[\Delta c_i] = 0, \ \text{var}[\Delta c_i] = \sigma_{c_i}^2, \ i = 1, 2, \cdots n$$

where Δa_{ij}, Δb_i and Δc_i are the random coefficients in ΔA, Δb, and Δc, respectively. To proceed with the general development, a general input signal is chosen. So that the input signal, $z(t)$, will excite all natural frequencies of the gyroscope equally, the input signal, $z(t)$, is chosen to be a zero mean, white noise signal with first and second order statistics

$$E\left[z(t)\right] = 0 \quad \text{and} \quad R_z(\tau) = \delta(\tau)$$

where $R_z(\tau)$ represents the autocorrelation function of $z(t)$.

Under these assumptions, $\Delta y(t)$ is the sum of $n^2 + 2n$ zero mean, mutually independent, random processes. In the Fourier transform domain, the total random variation, $\Delta y(t)$ can be written:

$$\Delta \mathbf{Y}(j\omega) = \sum_{i=1}^{n}\sum_{j=1}^{n} \Delta a_{ij} \mathbf{H}_{cbij}(j\omega)\mathbf{Z}(j\omega) + \sum_{i=1}^{n} \Delta b_i \mathbf{H}_{ci}(j\omega)\mathbf{Z}(j\omega) + \sum_{i=1}^{n} \Delta c_i \mathbf{H}_{bi}(j\omega)\mathbf{Z}(j\omega)$$

where $\mathbf{H}_{cbij}(j\omega)$, $\mathbf{H}_{ci}(j\omega)$, and $\mathbf{H}_{bi}(j\omega)$ are the Fourier transforms of the unit impulse responses, $h_{cbij}(t)$, $h_{ci}(t)$, and $h_{bi}(t)$, respectively. In this development, the Fourier transforms $\mathbf{H}_{cbij}(j\omega)$, $\mathbf{H}_{ci}(j\omega)$, and $\mathbf{H}_{bi}(j\omega)$ will be referred to as nominal subsystems and idealizations of the nominal subsystems, $\mathbf{H}_{cbij}(j\omega)$, $\mathbf{H}_{ci}(j\omega)$, and $\mathbf{H}_{bi}(j\omega)$, are used. The basic idealization approximates many systems whose frequency responses are lowpass filters with relatively constant amplitudes. Note that all $n^2 + 2n$ nominal subsystems have the same denominators (the same poles); therefore, the passbands of all the $n^2 + 2n$ subsystems will be approximated over the same frequency range of $-\omega_s \leq \omega \leq -\omega_s$; however, the subsystem gains are assumed to be different. Therefore, the idealized subsystem frequency responses will be modelled as ideal lowpass filters which can be defined as:

$$\mathbf{H}_{cbij}(j\omega) = \begin{cases} K_{cbij}e^{-j\omega} & |\omega| < \omega_s \\ 0 & |\omega| > \omega_s \end{cases}$$

$$\mathbf{H}_{ci}(j\omega) = \begin{cases} K_{ci}e^{-j\omega} & |\omega| < \omega_s \\ 0 & |\omega| > \omega_s \end{cases}$$

and $\mathbf{H}_{bi}(j\omega) = \begin{cases} K_{bi}e^{-j\omega} & |\omega| < \omega_s \\ 0 & |\omega| > \omega_s \end{cases}$

where K_{cbij}, K_{ci}, and K_{bi} are real constants. Because the $n^2 + 2n$ terms in $\Delta y(t)$ are mutually independent, $z(t)$ is a white noise, and the subsystems are band limited, the power spectral density (PSD), $S_{\Delta y}(\omega)$, of the random variation $\Delta y(t)$ is:

$$S_{\Delta y}(\omega) = \sum_{i=1}^{n}\sum_{j=1}^{n}\sigma_{\Delta a_{ij}}^2 \left|\mathbf{H}_{cbij}(j\omega)\right|^2 + \sum_{i=1}^{n}\sigma_{\Delta b_i}^2 \left|\mathbf{H}_{ci}(j\omega)\right|^2 + \sum_{i=1}^{n}\sigma_{\Delta c_i}^2 \left|\mathbf{H}_{bi}(j\omega)\right|^2 \quad -\omega_s < \omega < \omega_s$$

Using the idealized frequency response functions, this can be rewritten

$$S_{\Delta y}(\omega) = \sum_{i=1}^{n}\sum_{j=1}^{n}K_{cb_{ij}}^2\sigma_{\Delta a_{ij}}^2 + \sum_{i=1}^{n}K_{c_i}^2\sigma_{\Delta b_i}^2 + \sum_{i=1}^{n}K_{b_i}^2\sigma_{\Delta c_i}^2 \quad -\omega_s < \omega < \omega_s$$

The total power, $P_{\Delta y}$, in the random variation $\Delta y(t)$ is

$$P_{\Delta y} = \int_{\omega=-\infty}^{\omega=\infty} S_{\Delta y}(\omega)d\omega = \int_{\omega=-\omega_s}^{\omega=\omega_s} S_{\Delta y}(\omega)d\omega = \sum_{i=1}^{n}\sum_{j=1}^{n}2K_{cbij}^2\sigma_{\Delta a_{ij}}^2\omega_s + \sum_{i=1}^{n}2K_{bi}^2\sigma_{\Delta b_i}^2\omega_s + \sum_{i=1}^{n}2K_{ci}^2\sigma_{\Delta c_i}^2\omega_s$$

therefore $S_{\Delta y}(\omega)$ can be rewritten

$$S_{\Delta y}(\omega) = \begin{cases} \dfrac{P_{\Delta y}}{2\omega_s} & -\omega_s < \omega < \omega_s \\ 0 & \text{elsewhere} \end{cases} \tag{19}$$

Equation (19) indicates that the total power in the random variation $\Delta y(t)$ is uniformly distributed over the passband of the gyroscope.

5. TRANSFORMING $\Delta Y(T)$ INTO AN (ALMOST) WHITE NOISE

A dynamic system matching technique (DSMT) [Reference 4] can be used to transform the output noises in a set of real gyroscopes defined in Equations (13) and (14) into an (almost) white noise. The development in the previous section is now extended to a set of M real gyroscopes, each defined by Equations (13) and (14) and with a common input signal, $z(t)$, (which is assumed to be a white noise). The output signals of each real gyroscope are combined through a switching circuit. The switching circuit's output signal at a specific time t is the output signal, $y(t)$, of one of the M real gyroscopes which has been randomly selected (equal likelihood) by the switching circuit. The switching circuit continues to output that gyroscope's output signal over a fixed interval of time T, the *switching circuit period*, and after that time interval, T, the switching circuit randomly selects a different gyroscope and outputs its output signal over the next switching circuit period. The switching circuit continues to randomly select gyroscope outputs for each successive switching circuit period, *ad infinitum*. Note that if each gyroscope were a *nominal gyroscope*, each of their outputs would be identical and, assuming perfect switching, the output of the switching circuit would equal the output of the nominal gyroscope; however, with a set of M real systems, each containing coefficients with random perturbations, the switching circuit output changes randomly about the nominal output signal for each switching circuit period, T.

Let the coefficient matrix and vectors of the M real gyroscopes be denoted $\mathbf{A} + \Delta\mathbf{A}^i$, $\mathbf{b} + \Delta\mathbf{b}^i$, and $\mathbf{c} + \Delta\mathbf{c}^i$, for $i = 1, 2, \cdots, M$, where the Δ-quantities represent $n^2 + 2n$ random coefficient variations of the ith real gyroscope about the nominal coefficient values. It is assumed that over the sample space of all possible gyroscopes that each element of $\Delta\mathbf{A}$, $\Delta\mathbf{b}$, and $\Delta\mathbf{c}$ can be represented by a zero mean, finite variance random variable and that the elements of $\Delta\mathbf{A}$, $\Delta\mathbf{b}$, and $\Delta\mathbf{c}$ are mutually independent. For example, the first element, Δa_{11}, of the matrix, $\Delta\mathbf{A}$, is a random variable such that

$$E\left[\Delta a_{11}\right] = 0 \text{ and } \mathrm{var}\left[\Delta a_{11}\right] = \sigma^2_{\Delta a_{11}} < \infty.$$

If the set of M gyroscopes is chosen from the sample space of all possible gyroscopes, each of the coefficient variations for this set is a random sample selected from the sample space of the corresponding random variable. For example, the set of coefficient variations, Δa_{11}^i, $i = 1, 2, \cdots, M$, is a random sample taken from the sample space of the random variable Δa_{11}, and because the coefficient variations are mutually independent,

$$E\left[\Delta a_{11}\right] = 0, \ \mathrm{var}\left[\Delta a_{11}\right] = \sigma^2_{\Delta a_{11}}$$

and

$$E\left[\Delta a_{11}^i \Delta a_{11}^j\right] = 0, \text{ for } i \neq j.$$

Similar relationships hold for all the other $n^2 + 2n - 1$ coefficients as well. Because the switching circuit sequentially samples the random sample and the elements of the random sample are independent, then the sequence of switched outputs formed by the switching circuit can (heuristically) be looked upon as an 'almost white' sequence that becomes 'whiter' as M increases. Because the number of independent elements in the sequence is finite and each element will, with non-zero probability, be chosen more than once by the switch, time correlation will exist in these random sequences and they will only be white in the limit as M goes to infinity and T goes to zero. The rest of this section will develop analysis to validate these statements.

To show how a DSMT affects the output noise in the system due to random variations in the coefficients, PSDs and total powers of the noise in the DSMT output are evaluated. Consider a continuous-time function $f(t)$ generated by a time sequence of pulses, $p_k(t)$, $-\infty < k < \infty$ such that

$$f(t) = \sum_k p_k(t)$$

where each pulse $p_k(t)$ has width T, has unity amplitude for $kT \le t < (k+1)T$, and is zero elsewhere. Obviously, the function $f(t) = 1$ for all of t.

Let $\Delta y^1(t), \Delta y^2(t), \cdots, \Delta y^M(t)$ represent the random linear perturbations in the outputs of the M real systems, each of which can be described by (15) and (16). For each discrete time kT, define a discrete random variable i_k whose sample space consists of the integers $1, 2, \cdots, M$ with each element of the sample space being equally likely to occur. If the elements of the set of random variables, i_k, $-\infty < k < \infty$, are mutually independent, then the time sequence of random variables can represent the set of real systems randomly chosen by the switching circuit in the DSMT system. Therefore, the random noise, $U(t)$, in the output of the DSMT can then be written as

$$U(t) = \sum_k \Delta y^{i_k}(t) p_k(t)$$

where $\Delta y^{i_k}(t)$ is chosen randomly by the switching circuit at the beginning, time kT, of the kth switching period and the result in Equation (18) gives:

$$\Delta y^{i_k}(t) = \sum_{i=1}^{n}\sum_{j=1}^{n}\Delta a_{ij}^{i_k}\int_{\tau=0}^{\tau=\infty}h_{cbij}(\tau)z(t-\tau)d\tau + \sum_{i=1}^{n}\Delta b_{i}^{i_k}\int_{\tau=0}^{\tau=\infty}h_{ci}(\tau)z(t-\tau)d\tau + \sum_{i=1}^{n}\Delta c_{i}^{i_k}\int_{\tau=0}^{\tau=\infty}h_{bi}(\tau)z(t-\tau)d\tau$$

Now $U(t)$ can be written

$$U(t) = \sum_{i=1}^{n}\sum_{j=1}^{n}\left[\sum_{k}\Delta a_{ij}^{i_k}\int_{\tau=0}^{\tau=\infty}h_{cbij}(\tau)z(t-\tau)d\tau\, p_{k}(t)\right] + \sum_{i=1}^{n}\left[\sum_{k}\Delta b_{i}^{i_k}\int_{\tau=0}^{\tau=\infty}h_{ci}(\tau)z(t-\tau)d\tau\, p_{k}(t)\right]$$

$$+ \sum_{i=1}^{n}\left[\sum_{k}\Delta c_{i}^{i_k}\int_{\tau=0}^{\tau=\infty}h_{bi}(\tau)z(t-\tau)d\tau\, p_{k}(t)\right]$$

$$(20)$$

Note that $U(t)$ is the sum of $n^2 + 2n$ zero mean, independent, random processes and that each of the random processes has the same basic form; therefore, we need to operate on only one of these random processes, extrapolate the results from that one to all the others, and then combine all the results to obtain the desired result for $U(t)$. If $U_{\Delta a_{ij}}(t)$, $U_{\Delta b_i}(t)$, and $U_{\Delta c_i}(t)$ are defined as the components of $U(t)$ produced by the random variations Δa_{ij}, Δb_i, and Δc_i, respectively, then

$$U_{\Delta a_{ij}}(t) = \left[\int_{\tau=0}^{\tau=\infty}h_{cbij}(\tau)z(t-\tau)d\tau\right]\left[\sum_{k}\Delta a_{ij}^{i_k}p_{k}(t)\right]$$

$$= U'_{\Delta a_{ij}}(t)\cdot U''_{\Delta a_{ij}}(t)$$

$$U_{\Delta b_{i}}(t) = \left[\int_{\tau=0}^{\tau=\infty}h_{ci}(\tau)z(t-\tau)d\tau\right]\left[\sum_{k}\Delta b_{i}^{i_k}p_{k}(t)\right]$$

$$= U'_{\Delta b_{i}}(t)\cdot U''_{\Delta b_{i}}(t)$$

$$U_{\Delta c_i}(t) = \left[\int_{\tau=0}^{\tau=\infty} h_{bi}(\tau)z(t-\tau)d\tau \right]\left[\sum_k \Delta c_i^{i_k} p_k(t) \right]$$

$$= U'_{\Delta c_i}(t) \cdot U''_{\Delta c_i}(t)$$

Because the input signal, $z(t)$, is independent of $\Delta a_{ij}^{i_k}$, $\Delta b_i^{i_k}{}_i$ and $\Delta c_i^{i_k}{}_i$ for all k, then $U'_{\Delta a_{ij}}(t), U''_{\Delta a_{ij}}(t), U'_{\Delta b_i}(t), U''_{\Delta b_i}(t), U'_{\Delta c_i}(t)$ and $U''_{\Delta c_i}(t)$ are mutually independent random processes.

6. THE POWER SPECTRAL DENSITY (PSD)
OF A GENERAL NOISE TERM

In Equation (20), the $n^2 + 2n$ terms are mutually independent random processes and they all have the same mathematical form; therefore, in the following development, a general term, that is representative of all $n^2 + 2n$ terms, is defined as:

$$U_g(t) = \sum_k \Delta a^{i_k} \int_{\tau=0}^{\tau=\infty} h(\tau)z(t-\tau)d\tau\, p_k(t) = \left[\int_{\tau=0}^{\infty} h(\tau)z(t-\tau)d\tau \right]\left[\sum_k \Delta a^{i_k} \cdot p_k(t) \right] = U'(t) \cdot U''(t)$$

$$(21)$$

The PSD of this general term is developed and the resulting PSD is extrapolated to all of the terms in Equation (20).

The autocorrelation function and power spectral density (PSD) of the general term $U_g(t)$ are developed as follows. The autocorrelation function of $U_g(t)$ can be written as

$$R_{U_g}(\tau) = \mathrm{E}[U_g(t)U_g(t+\tau)] = \mathrm{E}[U'(t)U''(t)U'(t+\tau)U''(t+\tau)]$$
$$= \mathrm{E}[U'(t)U'(t+\tau)]\cdot \mathrm{E}[U''(t)U''(t+\tau)] = R_{U'}(\tau)\cdot R_{U''}(\tau)$$

The power spectral density (PSD) of $U_g(t)$ is given by:

$$S_{U_g}(\omega) = \int_{\tau=-\infty}^{\tau=\infty} e^{-j\omega\tau} R_{U'}(\tau) R_{U''}(\tau) d\tau$$

and by the complex convolution theorem

$$S_{U_g}(\omega) = \frac{1}{2\pi} \int_{\upsilon=-\infty}^{\upsilon=\infty} S_{U'}(\omega-\upsilon) S_{U''}(\upsilon) d\upsilon \qquad (22)$$

where $S_{U'}(\omega)$ is the PSD of $U'(t)$ and $S_{U''}(\omega)$ is the PSD of $U''(t)$. Recall from Section 4, that $z(t)$ is assumed to be white noise with first and second order statistics given by the mean and the autocorrelation function:

$$E[z(t)] = 0 \quad \text{and} \quad R_z(\tau) = \delta(\tau)$$

The autocorrelation function of $U'(t)$ is given by:

$$R_{U'}(\tau) = E[U'(t)U'(t+\tau)] = \int_{\theta=0}^{\theta=\infty} \int_{\alpha=0}^{\alpha=\infty} h(\theta)h(\alpha)R_z(\tau-\alpha+\theta)d\alpha d\theta$$

Given the white noise assumption for $z(t)$, this becomes:

$$R_{U'}(\tau) = \int_{\theta=0}^{\theta=\infty} h(\theta)h(\tau+\theta)d\theta$$

The PSD of $U'(t)$ is given by the Fourier transform of $R_{U'}(\tau)$, that is:

$$S_{U'}(\omega) = |H(j\omega)|^2$$

where $H(j\omega)$ is the Fourier transform of $h(t)$. As discussed in conjunction with Section 4, an idealization of $H(j\omega)$ is used. The idealized system used is the (two-sided) ideal low-pass filter defined by:

$$H(j\omega) = \begin{cases} Ke^{-j\omega} & |\omega_s| \leq \omega \\ 0 & |\omega_s| > \omega \end{cases}$$

$$= Ku(\omega_s - \omega)u(\omega_s + \omega)e^{-j\omega}$$

where $[-\omega_s, \omega_s]$ is the pass-band, where K is the constant amplitude over the pass-band, and where $u(\omega)$ is the unit step function defined as

$$u(\omega) = \begin{cases} 1 & \omega \geq 0 \\ 0 & \omega < 0 \end{cases}$$

The two-sided filter is used because the PSDs are all two-sided, that is, they are defined for negative frequencies and the system must deal with negative frequencies. Using this idealized system, the PSD of $U'(t)$ is given by:

$$S_{U'}(\omega) = \begin{cases} K^2 & |\omega_s| \leq \omega \\ 0 & |\omega_s| > \omega \end{cases}$$

$$= K^2 u(\omega_s - \omega)u(\omega_s + \omega)$$

and the PSD of $U_g(t)$ can be written:

$$S_{U_g}(\omega) = \frac{1}{2\pi} \int_{\upsilon=-\infty}^{\upsilon=\infty} K^2 u(\omega_s - \omega + \upsilon)u(\omega_s + \omega - \upsilon)S_{U''}(\upsilon)d\upsilon = \frac{K^2}{2\pi} \int_{\upsilon=\omega-\omega_s}^{\upsilon=\omega+\omega_s} S_{U''}(\upsilon)d\upsilon \quad -\infty < \omega < \infty$$

$$(23)$$

The final step in generating the PSD of $U_g(t)$ is to develop the autocorrelation function and PSD of the noise component

$$U''(t) = \sum_k \Delta a^{i_k} \cdot p_k(t) \qquad -\infty < t < \infty \tag{24}$$

The mean and variance of the noise component $U''(t)$ are easily shown to be:

$$E[U''(t)] = 0, \quad \text{var}[U''(t)] = \sigma_{\Delta a}^2 \qquad -\infty < t < \infty$$

The autocorrelation function of $U''(t)$ is developed as follows:

$$R_{U''}(t,\tau) = E[U''(t)U''(\tau)]$$

For $t,\tau \in [kT, (k+1)T)$, $U''(t) = U''(\tau) = \Delta a^{i_k}$, and $R_{U''}(t,\tau) = E[(\Delta a^{i_k})^2] = \sigma_{\Delta a}^2$

For $t \in [jT, (j+1)T)$, $\tau \in [kT, (k+1)T)$, $j \neq k$ either

$U''(t) = U''(\tau)$ with probability $\dfrac{1}{M}$ or $U''(t) \neq U''(\tau)$ with probability $\dfrac{M-1}{M}$

therefore, for $j \neq k$

$$R_{U''}(t,\tau) = E[U''(t)U''(\tau) \mid U''(t) = U''(\tau)]\Pr[U''(t) = U''(\tau)]$$
$$+ E[U''(t)U''(\tau) \mid U''(t) \neq U''(\tau)]\Pr[U''(t) \neq U''(\tau)]$$

$$= \frac{\sigma_{\Delta a}^2}{M} + E[U''(t)U''(\tau) \mid U''(t) \neq U''(\tau)]\left(\frac{M-1}{M}\right)$$

If $j \neq k$ and $U''(t) \neq U''(\tau)$, then $U''(t)$ and $U''(\tau)$ are independent and

$$E[U''(t)U''(\tau)\,|\,U''(t) \neq U''(\tau)] = 0,$$

therefore, $R_{U''}(t,\tau) = \dfrac{\sigma_{\Delta a}^2}{M}$ for $t \in [jT,(j+1)T)$, $\tau \in [kT,(k+1)T)$, $j \neq k$

Then, for all t and τ, the autocorrelation function can be written as

$$R_{U''}(t,\tau) = \left(\frac{M-1}{M}\right) \cdot \sigma_{\Delta a}^2 \cdot \delta(j,k) + \frac{\sigma_{\Delta a}^2}{M}$$

where $\delta(j,k)$ is a Kronecker delta, that is,

$$\delta(j,k) = \begin{cases} 1 & j = k, \text{that is, for } t,\tau \in [kT,(k+1)T) \\ 0 & j \neq k, \text{that is, for } t \in [jT,(j+1)T),\ \tau \in [kT,(k+1)T),\ j \neq k \end{cases}$$

Note that $R_{U''}(t,\tau)$ is not explicitly dependent on either t or τ. The power spectral density of $U''(t)$ is given by

$$S_{U''}(\omega) = \int_{\theta=-\infty}^{\theta=\infty} e^{-j\omega\theta} \left[\left(\frac{M-1}{M}\right) \cdot \sigma_{\Delta a}^2 \cdot \delta(j,k) + \frac{\sigma_{\Delta a}^2}{M}\right] d\theta$$

$$= \int_{\theta=-\infty}^{\theta=\infty} e^{-j\omega\theta} \left(\frac{M-1}{M}\right) \cdot \sigma_{\Delta a}^2 \cdot \delta(j,k)\,d\theta + \int_{\theta=-\infty}^{\theta=\infty} e^{-j\omega\theta} \frac{\sigma_{\Delta a}^2}{M}\,d\theta \qquad -\infty < \omega < \infty$$

The second term in $S_{U''}(\omega)$ is the Fourier transform of the constant $\dfrac{\sigma_{\Delta a}^2}{M}$ and is given by

$$\int_{\theta=-\infty}^{\theta=\infty} e^{-j\omega\theta} \frac{\sigma_{\Delta a}^2}{M}\,d\theta = 2\pi \frac{\sigma_{\Delta a}^2}{M} \delta(\omega)$$

where $\delta(\omega)$ is the Dirac delta function. This term indicates that the noise component $U''(t)$ contains a random DC offset. The first term in $S_{U''}(\omega)$ is given by

$$\int_{\theta=-\infty}^{\theta=\infty} e^{-j\omega\theta} \cdot \left(\frac{M-1}{M}\right) \cdot \sigma_{\Delta a}^2 \cdot \delta(j,k)d\theta = \left(\frac{M-1}{M}\right)\sigma_{\Delta a}^2 \int_{\theta=-\infty}^{\theta=\infty} e^{-j\omega\theta}\delta(j,k)d\theta$$

The integral $\displaystyle\int_{\theta=-\infty}^{\theta=\infty} e^{-j\omega\theta}\delta(j,k)d\theta$ is integrated as follows.

$$\int_{\theta=-\infty}^{\theta=\infty} e^{-j\omega\theta}\delta(j,k)d\theta = \int_{\theta=0}^{\theta=\frac{\sqrt{2T}}{2}} e^{-j\omega\theta}(\sqrt{2T}-2\theta)d\theta + \int_{\theta=-\frac{\sqrt{2T}}{2}}^{\theta=0} e^{-j\omega\theta}(\sqrt{2T}+2\theta)d\theta$$

$$= \sqrt{2T}\left[\int_{\theta=0}^{\theta=\frac{\sqrt{2T}}{2}} e^{-j\omega\theta}d\theta + \int_{\theta=-\frac{\sqrt{2T}}{2}}^{\theta=0} e^{-j\omega\theta}d\theta\right] - 2\left[\int_{\theta=0}^{\theta=\frac{\sqrt{2T}}{2}} \theta e^{-j\omega\theta}d\theta - \int_{\theta=-\frac{\sqrt{2T}}{2}}^{\theta=0} \theta e^{-j\omega\theta}d\theta\right]$$

The first term, after integration, becomes:

$$\sqrt{2T}\left[\int_{\theta=0}^{\theta=\frac{\sqrt{2T}}{2}} e^{-j\omega\theta}d\theta + \int_{\theta=-\frac{\sqrt{2T}}{2}}^{\theta=0} e^{-j\omega\theta}d\theta\right] = 2T^2\text{sinc}\left(\omega\frac{\sqrt{2T}}{2}\right)$$

where $\text{sinc}\, x = \dfrac{\sin x}{x}$. The second term after integration becomes:

$$-2\left[\int_{\theta=0}^{\theta=\frac{\sqrt{2T}}{2}} \theta e^{-j\omega\theta}d\theta - \int_{\theta=-\frac{\sqrt{2T}}{2}}^{\theta=0} \theta e^{-j\omega\theta}d\theta\right] = -2T^2\text{sinc}\left(\omega\frac{\sqrt{2T}}{2}\right) + T^2\text{sinc}^2\left(\omega\frac{\sqrt{2T}}{4}\right)$$

Finally then the two-sided power spectral density (PSD) of the noise component $U''(t)$ is given by

$$S_{U''}(\omega) = \left(\frac{M-1}{M}\right) \cdot \sigma_{\Delta a}^2 \cdot T^2 \mathrm{sinc}^2\left(\omega\frac{\sqrt{2}T}{4}\right) + 2\pi\left(\frac{\sigma_{\Delta a}^2}{M}\right)\delta(\omega) \quad -\infty < \omega < \infty$$

This can also be written

$$S_{U''}(\omega) = 4\left(\frac{M-1}{M}\right) \cdot \frac{\sigma_{\Delta a}^2}{\omega^2} \cdot \left[1 - \cos\frac{\omega T}{\sqrt{2}}\right] + 2\pi\left(\frac{\sigma_{\Delta a}^2}{M}\right)\delta(\omega) \quad -\infty < \omega < \infty \tag{25}$$

Using the second form for $S_{U''}(\omega)$ and Equation (23), the PSD of $U_g(t)$ is given by

$$S_{U_g}(\omega) = \frac{K^2}{2\pi} \int\limits_{\upsilon=\omega-\omega_s}^{\upsilon=\omega+\omega_s} S_{U''}(\upsilon)d\upsilon$$

$$= K^2\left(\frac{2}{\pi}\right)\left(\frac{M-1}{M}\right)\sigma_{\Delta a}^2 \int\limits_{\upsilon=\omega-\omega_s}^{\upsilon=\omega+\omega_s} \frac{1}{\upsilon^2}\left[1 - \cos\left(\frac{\upsilon T}{\sqrt{2}}\right)\right]d\upsilon + K^2\left(\frac{\sigma_{\Delta a}^2}{M}\right)\int\limits_{\upsilon=\omega-\omega_s}^{\upsilon=\omega+\omega_s}\delta(\upsilon)d\upsilon \;;\; -\infty < \omega < \infty \tag{26}$$

7. THE POWER SPECTRAL DENSITY OF THE TOTAL NOISE IN THE DSMT OUTPUT

Recalling that the total random noise, $U(t)$, in the output of the DSMT, is given by Equation (20) and that the PSD of each term in $U(t)$ has the form in Equation (26), then the PSD of $U(t)$ can be written as:

$$S_U(\omega) = \sum_{i=1}^{n}\sum_{j=1}^{n}\left[\left(\frac{K_{cb_{ij}}^2\sigma_{\Delta a_{ij}}^2}{M}\right)\left[\left(\frac{2(M-1)}{\pi}\right)\int\limits_{\upsilon=\omega-\omega_s}^{\upsilon=\omega+\omega_s}\frac{1}{\upsilon^2}\left[1-\cos\left(\frac{\upsilon T}{\sqrt{2}}\right)\right]d\upsilon + \int\limits_{\upsilon=\omega-\omega_s}^{\upsilon=\omega+\omega_s}\delta(\upsilon)d\upsilon\right]\right]$$

$$+ \sum_{i=1}^{n}\left[\left(\frac{K_{c_i}^2\sigma_{\Delta b_i}^2}{M}\right)\left[\left(\frac{2(M-1)}{\pi}\right)\int\limits_{\upsilon=\omega-\omega_s}^{\upsilon=\omega+\omega_s}\frac{1}{\upsilon^2}\left[1-\cos\left(\frac{\upsilon T}{\sqrt{2}}\right)\right]d\upsilon + \int\limits_{\upsilon=\omega-\omega_s}^{\upsilon=\omega+\omega_s}\delta(\upsilon)d\upsilon\right]\right]$$

$$+ \sum_{i=1}^{n}\left[\left(\frac{K_{b_i}^2\sigma_{\Delta c_i}^2}{M}\right)\left[\left(\frac{2(M-1)}{\pi}\right)\int\limits_{\upsilon=\omega-\omega_s}^{\upsilon=\omega+\omega_s}\frac{1}{\upsilon^2}\left[1-\cos\left(\frac{\upsilon T}{\sqrt{2}}\right)\right]d\upsilon + \int\limits_{\upsilon=\omega-\omega_s}^{\upsilon=\omega+\omega_s}\delta(\upsilon)d\upsilon\right]\right] \quad -\infty < \omega < \infty$$

Recalling from Section 4 that the total power in the random variation $\Delta y(t)$ of a single gyroscope is given by;

$$P_{\Delta y} = 2 \left[\sum_{i=1}^{n} \sum_{j=1}^{n} K_{\mathbf{cb}_{ij}}^2 \sigma_{\Delta a_{ij}}^2 + \sum_{i=1}^{n} K_{\mathbf{c}_i}^2 \sigma_{\Delta b_i}^2 + \sum_{i=1}^{n} K_{\mathbf{b}_i}^2 \sigma_{\Delta c_i}^2 \right] \omega_s$$

then $S_U(\omega)$ can be written in a more convenient form as:

$$S_U(\omega) = P_{\Delta y} \left[\left(\frac{(M-1)}{M} \right) \frac{1}{\pi \omega_s} \int_{\upsilon = \omega - \omega_s}^{\upsilon = \omega + \omega_s} \frac{1}{\upsilon^2} \left[1 - \cos\left(\frac{\upsilon T}{\sqrt{2}} \right) \right] d\upsilon + \frac{1}{2M\omega_s} \int_{\upsilon = \omega - \omega_s}^{\upsilon = \omega + \omega_s} \delta(\upsilon) d\upsilon \right] \quad -\infty < \omega < \infty \quad (27)$$

Recall from Section 4 (Equation (19) below), that the PSD of the random variation $\Delta y(t)$ of a single gyroscope is given by;

$$S_{\Delta y}(\omega) = \sum_{i=1}^{n} \sum_{j=1}^{n} K_{\mathbf{cb}_{ij}}^2 \sigma_{\Delta a_{ij}}^2 + \sum_{i=1}^{n} K_{\mathbf{c}_i}^2 \sigma_{\Delta b_i}^2 + \sum_{i=1}^{n} K_{\mathbf{b}_i}^2 \sigma_{\Delta c_i}^2 = \frac{P_{\Delta y}}{2\omega_s} \quad -\omega_s < \omega < \omega_s$$

that is, the PSD is uniformly distributed over the frequency interval $-\omega_s < \omega < \omega_s$. Note that the noise $U(t)$ in the output of a DSMT in which the outputs of M gyroscopes have been combined, is distributed over the infinite frequency interval $-\infty < \omega < \infty$, but it is apparently not a white noise because it is not uniformly distributed over the frequency interval. It will now be shown that in a limiting case, $U(t)$ does approach a white noise.

Consider an approximation for $S_U(\omega)$ as represented in Equation (27). It is assumed that M, the number of gyroscopes in the DSMT system, is very large and T, the switching period in the DSMT system, is very small. Note that the integral in the second term of Equation (27) is either zero or one (depending upon whether the interval of integration contains the origin); therefore the second term in Equation (27) becomes very small as

M becomes very large. In the first term of Equation (27), $\dfrac{(M-1)}{M} \cong 1$ for

large M and as T becomes very small, $\cos\left(\dfrac{\upsilon T}{\sqrt{2}}\right)$ can be approximated as

$\cos\left(\dfrac{\upsilon T}{\sqrt{2}}\right) \cong 1 - \dfrac{(\upsilon T)^2}{4}$. The final approximation for $S_U(\omega)$ becomes:

$$S_U(\omega) \cong P_{\Delta y}\left[\frac{1}{\pi\omega_s}\int_{\upsilon=\omega-\omega_s}^{\upsilon=\omega+\omega_s}\frac{T^2}{4}d\upsilon\right] = P_{\Delta y}\left[\frac{T^2}{2\pi}\right] \qquad -\infty < \omega < \infty \qquad (28)$$

Note that this approximate PSD is uniformly distributed over the infinite interval $-\infty < \omega < \infty$, and thus for very large M and very small T, $U(t)$ can be considered an 'almost' white noise. The DSMT system can now be represented by the state equation:

$$\frac{d\mathbf{x}(t)}{dt} = \mathbf{A}\mathbf{x}(t) + \mathbf{b}z(t) \qquad (29)$$

and the output equation:

$$y(t) = \mathbf{c}^T\mathbf{x}(t) + U(t) \qquad (30)$$

where $U(t)$ is an 'almost' white noise. The analysis in Section 3 is now applicable to a set of these DSMT systems, that is, the outputs of a set of N DSMT systems, each containing M gyroscopes, are averaged and the average output drives a standard Kalman filter.

8. TRANSFORMING A RANDOM BIAS INTO AN 'ALMOST' WHITE NOISE

The development in this section follows the initial part of the development in Section 5. Consider a set of M gyroscopes where each element has the same state equation:

$$\frac{d\mathbf{x}(t)}{dt} = \mathbf{A}\mathbf{x}(t) + \mathbf{b}z(t) \tag{31}$$

but a different output equation:

$$y(t) + \Delta y_d^i(t) = \mathbf{c}^T \mathbf{x}(t) + d^i \; ; \; i = 1, 2, \cdots M \tag{32}$$

where d^i; $i = 1, 2, \cdots, M$ are a set of mutually independent zero mean scalar random biases with a common variance σ_d^2. $y(t)$ is the true output of the gyroscope generated by the input $z(t)$ and $\Delta y_d^i(t)$ is the measurement error generated by the random bias d^i. Obviously, $\Delta y_d^i(t) = d^i$.

Assume that the outputs of a set of M gyroscopes are combined through a DSMT system as described in Section 5. Consider a continuous-time function $f(t)$ generated by a time sequence of pulses, $p_k(t)$, $-\infty < k < \infty$ such that

$$f(t) = \sum_k p_k(t)$$

where each pulse $p_k(t)$ has width T, has unity amplitude for $kT \le t < (k+1)T$, and is zero elsewhere. Obviously, the function $f(t) = 1$ for all of t. Let $\Delta y_d^1(t) = d^1, \Delta y_d^2(t) = d^2, \cdots, \Delta y_d^M(t) = d^M$ represent the random biases in the outputs of the M gyroscopes. For each discrete time kT, define a discrete random variable i_k whose sample space consists of the integers $1, 2, \cdots, M$ with each element of the sample space being equally likely to occur. If the elements of the set of random variables, i_k, $-\infty < k < \infty$, are mutually independent, then the time sequence of random variables can represent the set of gyroscopes randomly chosen by the switching circuit in the DSMT system. Therefore, the random noise, $D(t)$, in the output of the DSMT can then be written as

$$D(t) = \sum_k d^{i_k} \cdot p_k(t)$$

where d^{i_k} is chosen randomly by the switching circuit at the beginning, time kT, of the kth switching period kT. Note that $D(t)$ has the same form as $U''(t)$ in Equation (24); therefore the PSD of $D(t)$ has the same form as the PSD for $U''(t)$ as given in Equation (25), that is, the PSD of $D(t)$ is given by:

$$S_D(\omega) = 4\left(\frac{M-1}{M}\right) \cdot \frac{\sigma_d^2}{\omega^2} \cdot \left[1 - \cos\frac{\omega T}{\sqrt{2}}\right] + 2\pi\left(\frac{\sigma_d^2}{M}\right)\delta(\omega) \quad -\infty < \omega < \infty \qquad (33)$$

Following the arguments at the end of Section 7, for very large M and very small T, the PSD of $D(t)$ can be approximated by:

$$S_D(\omega) = \sigma_d^2 \cdot T^2 \quad -\infty < \omega < \infty \qquad (34)$$

Apparently, the power in the random biases is distributed over the infinite frequency interval $-\infty < \omega < \infty$ and $D(t)$ is an 'almost' white noise.

9. COMBINING THE SIGNAL PROCESSING TECHNIQUES

In this section, the DSMT technique described in Section 5 is combined with the white noise processing technique described in Section 3. Consider a set of $N \cdot M$ real gyroscopes, each defined by a state equation:

$$\frac{d[\mathbf{x}(t) + \Delta\mathbf{x}(t)]}{dt} = (\mathbf{A} + \Delta\mathbf{A})[\mathbf{x}(t) + \Delta\mathbf{x}(t)] + (\mathbf{b} + \Delta\mathbf{b})z(t) \qquad (35)$$

and an output equation given by:

$$(y(t) + \Delta y(t)) = (\mathbf{c} + \Delta \mathbf{c})^T (\mathbf{x}(t) + \Delta \mathbf{x}(t)) + v(t) + d \tag{36}$$

where $\Delta \mathbf{A}, \Delta \mathbf{b}, \Delta \mathbf{c}$ random coefficient perturbations as defined in conjunction with Equations (15 and (16) and where $v(t)$ and d are a white noise and a random bias, respectively, with first and second order statistics as defined in conjunction with Equations (8) and (32), respectively. $\Delta \mathbf{A}, \Delta \mathbf{b}, \Delta \mathbf{c}, v(t)$ and d are assumed to be mutually independent for any single gyroscope and also between all $N \cdot M$ gyroscopes. The input $z(t)$ is assumed to be common to all $N \cdot M$ gyroscopes and, if all of the noise sources were zero, the output $y(t)$ would be the true output and be common to all $N \cdot M$ gyroscopes.

The set of $N \cdot M$ gyroscopes is divided into N subsets of M gyroscopes. The elements of the $i^{th}, i = 1, 2, \cdots, N$, subset of gyroscopes are combined in a DSMT system thus producing a single output $y(t) + \Delta y_i(t)$, where $y(t)$ is the true output and $\Delta y_i(t)$ is the error in the output due to the various noise sources in the ith subset. Each DSMT system can be represented by the state equation:

$$\frac{d\mathbf{x}(t)}{dt} = \mathbf{A}\mathbf{x}(t) + \mathbf{b}z(t) \tag{37}$$

and an output equation:

$$y_i(t) = \mathbf{c}^T \mathbf{x}(t) + v_i(t) + U_i(t) + D_i(t) \tag{38}$$

where $i = 1, 2, \cdots, N$. $v_i(t)$ is a white noise generated in the DSMT system by the switching process. This white noise is probabilistically the same as the white noise in any single gyroscope because switching between the various white noises in the subset, all with the same statistics, generates another white noise with the same statistics. $U_i(t)$ is an 'almost" white noise generated by the DSMT from the random coefficient variations $\Delta\mathbf{A}, \Delta\mathbf{b}, \Delta\mathbf{c}$ and $D_i(t)$ is an "almost" white noise generated from the random bias d. Assuming M to be very large and the switching period T in the DSMT system to be very small, all three noises in the DSMT output are considered to be white. The outputs of the N DSMT systems are now averaged as in Section 3, Equation (11). The final single equivalent gyroscope is given by the state equation;

$$\frac{d\mathbf{x}(t)}{dt} = \mathbf{A}\mathbf{x}(t) + \mathbf{b}z(t) \tag{39}$$

Similarly, an arithmetic average is taken of the set of measurement equations (8) generating the equation

$$\bar{y}(t) = \frac{1}{N}\left(\sum_{i=1}^{N} y_i(t)\right) = \mathbf{c}^T\left(\frac{1}{N}\sum_{i=1}^{N}\mathbf{x}_i(t)\right) + \frac{1}{N}\left(\sum_{i=1}^{N} v_i(t)\right) + \frac{1}{N}\left(\sum_{i=1}^{N} U_i(t)\right) + \frac{1}{N}\left(\sum_{i=1}^{N} D_i(t)\right)$$

which is rewritten

$$\bar{y}(t) = \mathbf{c}^T\mathbf{x}(t) + \bar{v}(t) + \bar{U}(t) + \bar{D}(t) \tag{40}$$

Estimates of the state $\mathbf{x}(t)$ in Equation (39) and the true output $y(t)$ of the gyroscopes generated by $z(t)$ can now be generated by applying a standard Kalman filter to Equations (39) and (40). Note that the second order statistics of all of the noises is reduced by a factor of N by the averaging process. The critical design parameters in this process are N, the number of DSMT systems, M, the number of elements in each DSMT system, and T, the switching period in DSMT system. Obviously, M and T can be different for each of the N DSMT systems.

CONCLUSION

Since microelectromechanical (MEMS) gyroscopes are small, lightweight, and inexpensive, they have many potential applications. Often, however, their accuracy is not adequate for many applications. In such cases, it may be appropriate to apply signal processing techniques to improve their accuracy. This chapter presents a signal processing technique for combining the scalar outputs of multiple gyroscopes, all measuring the same quantity, into a single output which provides a much more accurate (probabilistically) measurement of the quantity than a single gyroscope This technique is applicable if the gyroscopes are individually inexpensive and moderately accurate.

The general technique is composed of two techniques, each of which combines multiple output signals from a set of gyroscopes. The first technique for combining the measurements from multiple MEMS gyroscopes applies arithmetic averaging to the outputs of a set of gyroscopes, each of which has only a white noise corrupting its output. The averaging process has the effect of reducing the second order statistic of the white noise by a factor equal to the number of gyroscope outputs being averaged.

The second signal processing technique is applicable to gyroscopes whose outputs are corrupted by a random bias in gyroscope output and noise in the output due to variations in the gyroscope parameters due to the manufacturing process. The signal processing technique for this noise model is called a dynamic system matching technique (DSMT) and is used for combining the outputs of multiple MEMS gyroscopes in such a way that the random biases in the gyroscope outputs and the noises due to random variations in the system parameters are turned into (almost) white noises in the single combined gyroscope output. This technique has no (probabilistic) effect on any white noise that might also exist in the gyroscope output.

The final step in the development of the overall technique is to combine the two signal processing techniques. A set of many MEMS gyroscopes is chosen and divided into several subsets of gyroscopes. The outputs of the gyroscopes in each subset are combined using the Dynamic System Matching Technique resulting in a single gyroscope output corrupted by only an (almost) white noise. The single outputs from each subset are then combined by the averaging technique discussed in the first signal processing technique that reduces the white noise covariances in the resulting equivalent single gyroscope output. This single output is finally processed through a standard Kalman filter resulting in an output which statistically is much better than a single gyroscope. Note that the amount of computation in the Kalman filter is minimum.

In order for this technique to generate an accurate estimate of the true gyroscope output, a very large number of gyroscopes should be combined, thus, in a practical sense, the technique is applicable to inexpensive and moderately accurate gyroscopes.

REFERENCES

[1] Stubberud, A. R. and X.-H. Yu, "Signal Processing for Micro-Inertial Sensors," *RTO Meeting Proceedings 44, Advances in Vehicle Systems Concepts and Integration*, NATO Research and Technology Organization, Neuilly-Sur-Seine, France, April, 2000, pp. B11-1--B11-7.

[2] Stubberud, P. A. and Stubberud, A. R. (2008). A Signal Processing Technique for Improving the Accuracy of MEMS Inertial Sensors. *Proceedings of the Nineteenth International Conference on Systems Engineering*, pp. 13-18.

[3] Stubberud, P. A. and Stubberud, A.R. (2009). A Dynamic Element Matching Technique for Improving the Accuracy of MEMS Gyroscopes. *Proceedings of the Twentieth International Conference on Systems Engineering*, pp. 418-422.

[4] Stubberud, P. A., Stubberud, S. C. and Stubberud, A. R., A Dynamic System Matching Technique-A State Space Approach, *Proceedings of the 26th International Conference on Systems Engineering*, University of Nevada, Las Vegas, Las Vegas, August, 2017.

In: Gyroscopes
Editor: Marcel Gerste

ISBN: 978-1-53615-856-4
© 2019 Nova Science Publishers, Inc.

Chapter 3

MICROOPTICAL GYROS ON THE BASE OF PASSIVE RING RESONATORS

Yurii V. Filatov, Alexander S. Kukaev,
Egor V. Shalymov and Vladimir Yu. Venediktov

St. Petersburg State Electrotechnical University, St. Petersburg, Russia

ABSTRACT

Currently, optical gyroscopes are widely used in inertial navigation systems. At the same time, there is a pronounced tendency towards miniaturization, which spreads also to the navigation systems. The modern economics requires the development of mini and micro sensor devices to control highly dynamic technical systems, mobile robots, microsatellites, etc. The size of the controlled objects is continuously reduced, which in turn requires further miniaturization of gyroscopes. In response to this challenge during the last decades two directions in reducing the dimensions of optical gyroscopes were developed: one of which is aimed at creating compact active gyroscopes, and the second – compact passive gyroscopes. This chapter is devoted to the analysis of the main ways of creating and the development trends of compact passive optical gyroscopes. Their sizes are comparable to micromechanical (MEMS) gyroscopes sizes, and their ultimate sensitivity significantly exceeds the sensitivity of micromechanics and approaches the sensitivity of laser and fiber gyroscopes. Often such gyroscopes are called micro-optical (MOG) based on the size of their basic elements. Research on this topic was pushed forward by the development of integrated optics, and most of the prototypes of compact passive optical gyroscopes are built on the basis of optical integrated circuits, so these gyros are frequently called integrated optical gyros. Although the Sagnac effect is the basis of the operation of all optical gyroscopes, the method for determining the angular velocity depends on the connection circuit of their sensitive element. In the framework of this chapter compact passive optical gyroscopes are considered, first of all, from the point of its sensitive element connection circuit. In addition, various possible types of MOG sensitive elements are discussed since the characteristics of gyroscopes and optimal design are associated with their choice.

HISTORY REFERENCE

The principle of operation of all currently known optical gyroscopes (LG, FOG and MOG) is based on the Sagnac effect [1]. This effect was first demonstrated in 1913 by Georges Sagnac, who completed series of experiments devoted to the study of "ether" [2]. The scientist used an interferometer similar to that shown in Figure 1.

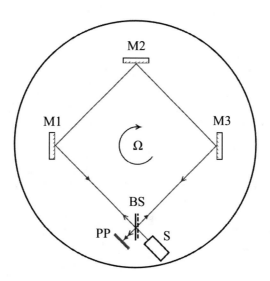

Figure 1. Scheme of the Sagnac interferometer.

In a ring interferometer consisting of several mirrors M1-M3 and a beam splitter BS, collimated and polarized light from source S was introduced. With the help of a beam splitter, the light beam from the source S was divided in two. Light bypassed the interferometer in mutually opposite directions, as shown in Figure 1. Then, both parts of the beam were combined by a beam splitter, interfered and sent to the photoplate PP. The interferometer was placed on a moving base. From experiment to experiment Georges Sagnac changed the angular velocity of the interferometer Ω and recorded photographs of the two-beam interference of counterpropagating waves (interference fringes) using photoplates. When studying the recorded interference patterns, the scientist discovered that the rotation of the interferometer leads to a shift of the interference fringes. A shift arises, since during a bypass of a rotating contour, the counterpropagating waves experience a nonreciprocal phase shift.

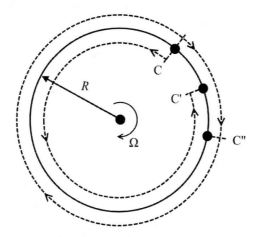

Figure 2. Propagation of counterpropagating waves along a rotating optical contour.

To understand what caused the phase shift, let us consider the propagation of two oppositely directed light waves in an optical contour in the shape of a circle with radius R (see Figure 2), rotating around a geometric center. Two waves begin to propagate simultaneously from the starting point C. Let a wave propagating counterclockwise (CCW) move in the direction opposite to the rotation of the optical contour. It meets the starting point in position C' shifted along the contour by some value l_1. The second wave propagating clockwise (CW) moves in the direction coinciding with the rotation of the optical contour. It meets the starting point in position C" displaced along the contour by some value l_2. Based on Figure 2, it is obvious that:

$$\begin{cases} t_{ccw} = \dfrac{2\pi R - l_1}{c} = \dfrac{2\pi R - t_{ccw}R\Omega}{c} \\ t_{cw} = \dfrac{2\pi R + l_2}{c} = \dfrac{2\pi R + t_{cw}R\Omega}{c} \end{cases}. \tag{1}$$

Here t_{cw} и t_{ccw} are the times of the optical path around the waves in the CW and CWW direction, respectively. Using expression (1), we can get:

$$\begin{cases} t_{ccw} = \dfrac{2\pi R}{c + R\Omega} \\ t_{cw} = \dfrac{2\pi R}{c - R\Omega} \end{cases} \tag{2}$$

From (2), it is obvious that the difference in time for the waves to bypass the circuit CW and CCW is equal to:

$$\Delta t = t_{cw} - t_{ccw} = \left(\frac{4\pi R^2}{c^2 + R^2\Omega^2} \right) \Omega. \tag{3}$$

It should be noted that in practice the condition $\Omega R \ll c$ is usually satisfied, therefore, expression (3) is usually written as follows:

$$\Delta t = \left(\frac{4S}{c^2} \right) \Omega. \tag{4}$$

Here S is the area covered by the optical contour in the plane perpendicular to the rotation axis. The latter expression is valid for the closed optical contour of any arbitrary shape [2]. Using the expression (4) one can easily determine the difference of the phases of the counterpropagating waves caused by the rotation while traversing the optical circuit (Sagnac phase shift)

$$\varphi_S = \left(\frac{8\pi S}{\lambda c} \right) \Omega , \qquad (5)$$

as well as the difference of eigenfrequencies of the ring resonator caused by the Sagnac effect for opposite directions of its bypass

$$\nu_S = \left(\frac{4S}{\lambda L} \right) \Omega . \qquad (6)$$

Here λ is the light wavelength in vacuum and L is the length of the optical contour perimeter.[1]

Thus, analyzing the interference pattern of the Sagnac interferometer allows estimation of the opposite waves phase shifts difference and, furthermore, the angular velocity can be determined using expression (5). However, the sensitivity of this measurement method is relatively small. One of the ways to increase sensitivity is to increase the area covered by the optical path. This idea is realized in fiber-optic gyroscopes (FOG), in which the optical contour of the interferometer is formed by a coil of an optical fiber with N turns. In this case, the area S is increased N times [3]. Another way to increase sensitivity is to move from phase to frequency measurements. This idea is realized in laser gyroscopes (LG). The angular velocity in LG is measured by the difference in the frequencies of the

---

[1] We explain here the sense and origin of the formulae (5) and (6) in a simplified manner. Rigorous theoretical description of the Sagnac effect is possible only within the general relativity theory (see, for instance, [2]). However, the very formulae (5) and (6) are correct.

opposing waves generated by a ring laser [3]. In this case, the expression (6) is used.

Both LG and FOG are currently characterized by fairly high accuracy. At the same time, they have a number of important advantages compared with other types of gyroscopes (primarily with mechanical ones): a large dynamic range of measured speeds; insensitivity to linear acceleration and to vibration; short starting up time; low power consumption; lack of cross-links when building a three-axis block; lack of bearings and of moving mechanical elements. Therefore, at present optical gyroscopes are widely used in inertial navigation systems. However, from year to year, the size of controlled objects continuously decreases, which in turn requires more and more compact gyroscopes. The size of optical gyroscopes is limited by the size of their sensitive element. During the last decades, two main directions have been developed to reduce the dimensions of optical gyroscopes, one of which is aimed at creating compact active gyroscopes (using micro-optical ring lasers), and the second - compact passive gyroscopes.

In the first case, a compact ring laser can be obtained using stimulated Brillouin scattering [4-6] or by activating the core of the waveguide or fiber with rare-earth elements [7-9]. The main obstacle in creating a MOG with an active resonator is the capture zone (locking of CW and CCW oscillations), which grows rapidly with decreasing perimeter. Another obstacle on the way to creating a MOG with an active resonator is the large width of the gain line of a compact ring laser, which leads to problems that can be difficult to get rid of when the resonator is miniaturized: multimode lasing, instability of bi-directional oscillations, etc. Therefore, the option of reducing the dimensions of an optical gyroscope through the use of passive sensitive elements seems more promising.

In the second case, either a multi-turn spiral waveguide coil, which forms a two-beam interferometer similar to the Sagnac interferometer, or a passive ring resonator (cavity, PRC), acts as a sensitive MOG element. Note that most of the research devoted to MOG, as well as the very first work on this subject are related to gyroscopes on the PRC. For the first time, a gyro based on PRC was proposed and studied in the late 70s. It utilized a number of mirrors and had dimensions that practically did not differ from the dimensions of LG (usually triangular or square configuration with a perimeter of some 20-60 cm). The main goal of building of the first PRCs was to create an optical gyroscope with sensitivity similar to that of LG, but with no capture zone (the zone corresponding to small angular velocities, in which LG is fundamentally insensitive). Initially it was considered that in the case of using PRC, synchronization of counterpropagating waves, causing a capture zone from LG, is fundamentally not possible due to the lack of generation in the resonator. However, during first experimental studies the capture zone in the PRC was discovered. Therefore, the macroscopic model of the PRC is not widely used. A few years later it turned out that the reason for the presence of frequency synchronization in the PRC are nonlinear effects that occur when using feedback circuits. With proper design of feedback circuits, the capture zone can be avoided. The development of integrated optics has created the basis for work aimed at significant reducing of cost and dimensions of optical gyroscopes. In this case, PRC is usually considered as the most promising sensitive element of MOG.

CONNECTION DIAGRAMS OF THE SENSITIVE ELEMENT OF PASSIVE MICRO-OPTICAL GYROSCOPES

Although the operation of all optical gyros, including MOGs, is based on the Sagnac effect, the method of determining the angular velocity in each specific MOG depends on the connection circuit of the sensitive element. From the point of view of the sensing element connection circuit all currently developed passive MOGs can be divided, as shown in Figure 3, into two types: resonator MOGs and interferometric MOGs.

Most of the experimentally implemented MOG prototypes are of the resonator type [10]. Various types of PRC are used as sensitive element of this type of MOG,. When PRC rotates due to the Sagnac effect, its resonant (eigen) frequencies shift. The direction of shift for opposite directions of bypass of the resonator is different and is determined by the direction of rotation of the PRC. The difference between the natural frequencies PRC for the opposite directions of its bypass is proportional to the angular velocity (see expression (6)). The operating principle of the resonator MOG is based on the determination of the difference between the natural frequencies of PRC, the magnitude of which determines the angular velocity. However, depending on the PRC connection scheme, various ways of determining the difference in the natural frequencies of the PRC are possible.

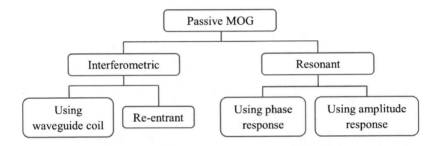

Figure 3. Classification of passive micro-optical gyroscopes according to the connection scheme of a sensitive element.

Resonator MOGs Using PRC Amplitude Response

As a rule, the PRC eigenfrequencies for opposite directions of its bypass are determined using only its amplitude characteristic. It is known that the PRC spectrum has extremes (dips – when PRC operates on reflection; peaks – when PRC operates on transmission) at eigenfrequencies and is similar to the spectrum of the Fabry-Perot interferometer [11]. Thus, by scanning the PRC in frequency and tracking the shifts of the extremes of its spectrum, you can determine the difference between the eigenfrequencies of the PRC and the corresponding angular velocity.

Let us consider in detail the operating principle of the resonant MOG using the amplitude characteristic of the PRC. Usually, resonator MOGs are implemented in a waveguide version (all optical elements are made in the form of one or several optical integrated circuits) with a PRC operating on reflection (see Figure 4.). The radiation from the source of monochromatic light 1 is divided by a splitter 2 (usually a Y-shaped splitter is used) into two channels. Since the width of the emission line of the light source must be significantly narrower than the width of the PRC spectral line, a laser is used as the light source in the resonator gyroscope. The frequency of emission of waves in both channels changes with the passage of the systems of modulators 3 and 4, which allows scanning the PRC spectrum. Depending on the frequency modulation law used in a particular MOG, modulator systems 3 and 4 can be composed of one or more series-connected phase and / or frequency modulators [12-15]. Using a directional coupler 9, the light from both channels is partially introduced into the PRC 10 (for example, a circular waveguide with a diameter of several centimeters) in mutually opposite directions (clockwise and counterclockwise – CW and CCW, respectively). The light that remains in the channels interferes with the set of waves that returned to the coupler 9 and bypassed the PRC optical circuit several times (there is a multipath interference). Then, through directional couplers 7 and 8, the radiation enters, respectively, photodetectors 5 and 6. Signals from both photodetectors are transmitted to the computing system 11.

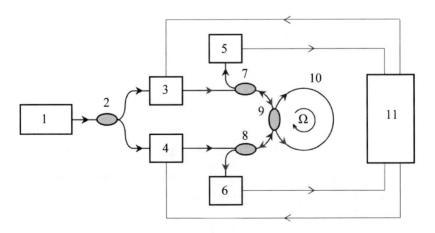

Figure 4. The operating principle of operation of the resonator micro-optical gyroscope using the amplitude characteristic of a passive ring resonator.

When a PRC is stationary relative to the inertial space, its natural frequencies for mutually opposite directions of the circuit around its contour are the same. The dependencies of the radiation intensity on photodetectors 5 and 6 (I_5 and I_6, respectively) on the frequency coincide and correspond to the graph in Figure 5 [11].

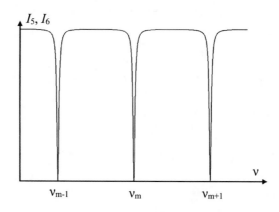

Figure 5. The dependence of the radiation intensity on the frequency at the outputs of a stationary relatively inertial space of a passive ring resonator.

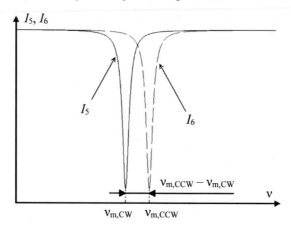

Figure 6. Splitting of the amplitude response of PRC with clockwise rotation.

If the frequency of the laser radiation is not equal to the frequency of one of the PRC modes v_m, then the light is almost not absorbed by the resonator and reaches the photodetectors. If the radiation frequency coincides with the eigenfrequency, then the light is retained in the resonator and dissipates in it. Thus, the minima of the PRC amplitude response (intensity at the photodetectors) corresponds to the eigenfrequencies of PRC v_m. When rotating MOG in the plane of Figure 4 (say, clockwise), the PRC eigenfrequencies for mutually opposite directions of its bypass are shifted in opposite directions due to the Sagnac effect (see Figure 6). Caused by the Sagnac effect, the PRC eigenfrequency difference for the opposite directions of the bypass of the resonator is determined by expression (6).

The computing system 11 controls the modulator systems 3 and 4 and performs PRC frequency scans. By changing the PRC amplitude response of (in particular, by the minima of the signals from photodetectors 5 and 6), the computing system determines the PRC resonant frequencies and adjusts the wave frequencies with the modulator systems, and also calculates its angular velocity by the difference of the natural frequencies PRC.

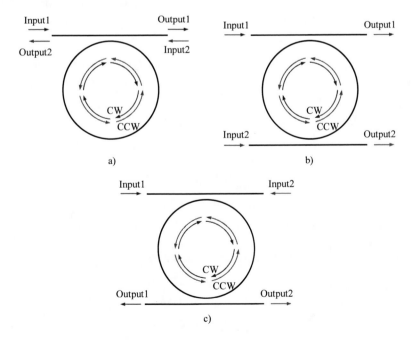

Figure 7. Input and output circuits, used in resonator micro-optical gyroscopes.

In the resonator gyroscope scheme discussed above, one optical communication loop (one auxiliary waveguide connected to the resonator) is used to input and output radiation from a passive ring resonator. There are other configurations of resonator MOG based on PRC, differing in the scheme of input and output of optical radiation from the resonator. However, in almost all resonant MOGs developed to date, one (see Figure 7a) or two (see Figure 7b, c) optical communication loops are used; they are usually based on directional couplers, for inputting and outputting radiation from the resonator.

The configuration, shown in Figure 7b, is a PRC with two loops of optical coupling between the resonator and auxiliary waveguides, one of which is used for input and output of radiation bypassing the resonator clockwise, and the other – for input and output of radiation bypassing the resonator counterclockwise [15]. If in the resonator MOG PRC is connected as shown in Figure 7a or b, the PRC is said to be reflective, and the corresponding MOG configurations are called reflective. There are also configurations in which one loop of the optical communication PRC is used to input radiation into the resonator, and the other – to output it (see Figure 7c) [12]. They are called transmissive. It is worth noting that reflective MOG configurations are more commonly used. This is due to the fact that multipath interferometers (including the PRC) working on reflection are structurally simpler and more sensitive to changes in the optical length of the resonator circuit (and therefore to the Sagnac effect) than those working in transmission mode.

In addition to those described above, there is another separate subtype of resonant MOGs using the amplitude characteristic of the PRC – MOG with compensation for the losses in the PRC. It is known that the higher is the quality factor (Q-factor) of the PRC, the higher is the sensitivity of the gyroscope based on it [3]. The quality factor can be improved by compensating for losses in the PRC. In this case, since the operation of the resonator is maintained below the generation threshold (losses are compensated, but generation is not observed), it remains passive. In one of the first MOG schemes with compensation for PRC losses it was proposed to symmetrically place two separate semiconductor optical amplifiers A_1

and A₂ in the waveguide ring resonator (see Figure 8) [16]. The use of semiconductor optical amplifiers makes it possible to compensate for the losses in resonator and to increase its Q-factor by several orders of magnitude. However, this scheme has a significant drawback. When a separate semiconductor amplifier is placed into the PRC, a material boundary appears at its edges and a portion of the radiation is reflected back. As a result, a part of the energy of the wave that goes around the PRC goes clockwise to the wave that goes around it counterclockwise, and vice versa. When determining the angular velocity, this causes an additional error [3].

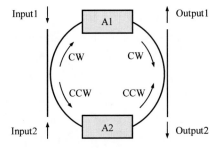

Figure 8. Passive ring resonator with integrated semiconductor optical amplifiers.

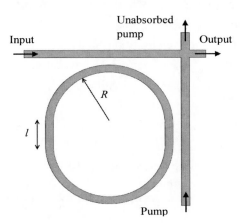

Figure 9. Passive ring resonator with uniformly distributed loss compensation.

To avoid this, another MOG configuration is used with a loss compensated PRC, in which active substance atoms implanted into the resonator play the role of an optical amplifier. A detailed theoretical and experimental study of one of the variants of such a PRC loss compensation scheme is given in [17]. A PRC made of silicate glass doped with two percent of neodymium oxide was investigated. The waveguide structure of the PRC consisted of a closed waveguide optically coupled to two auxiliary waveguides, as shown in Figure 9. Through one auxiliary waveguide, the measuring signal was introduced and output from the PRC, and through the other pump radiation was delivered to the PRC. The resonator was composed of two half-rings with radii $R = 8$ mm, connected by straight waveguide segments with a length l of about 3 mm. The resonator operated at a wavelength of 1.02 μm. A 150 mW semiconductor laser operating at a wavelength of 0.83 μm was used for pumping. Due to such compensation of losses, it was possible to increase the Q-factor of the resonator by more than 20 times (without pumping 8.32×10^5, while with pumping 1.89×10^7) [17]. There are other studies devoted to the two described loss compensation schemes PRC [18-21], but in all experiments in this area, the quality of only those PRCs that initially had a relatively low quality factor was increased by compensating for losses. At the same time, the quality factor has been raised to a level comparable to the quality of the best PRCs of the same size and type, but not higher. This is due to the fact that, firstly, the manufacturing technology of most high-quality PRCs does not allow for compensation of losses in them, and secondly, the introduction of elements into the sensor design that allows compensation of PRC losses (amplifiers in the PRC circuit etc.) increases the initial (before compensation) losses of the PRC. It is worth noting that the disadvantage of any loss compensation scheme is the partial "leakage" of the pump radiation into the measurement channel, which increases the noise at the photodetector. Also, the implementation of PRC loss compensation leads to an increase in MOG dimensions (since the pump source takes up additional space), power consumption and complicates the MOG design, which leads to an increase in cost and a decrease in MOG reliability. And since MOG is focused on using it in the low-cost and moderate sensitivity orientation and navigation systems for a wide range of applications, the disadvantages listed above are significant. Therefore, for the time being, from a practical point of view, resonator MOGs with loss compensation of PRCs are not of particular interest. However, in the long term, with the development of integrated optics technology, the disadvantages of this type of MOGs can be smoothed out and the application of the PRC loss compensation can be justified.

Resonator MOG Using Phase Response PRC

As noted earlier, the majority of MOG prototypes developed to date belong to the resonator type and expect the use of only the amplitude response of the PRC to determine the difference in natural frequencies proportional to the angular velocity. However, when PRC is rotated, due to the Sagnac effect not only its amplitude, but also the PRC phase characteristic splits. Figure 10 shows the form of the phase response of the PRC and its splitting for counterpropagating waves during the clockwise rotation of the PRC. In this case, as in the case of the amplitude characteristic, the magnitude of the splitting is determined by expression (6) and is proportional to the angular velocity.

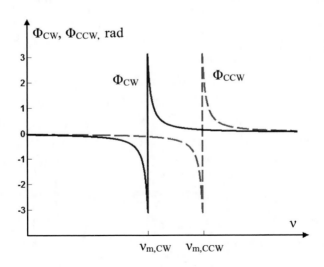

Figure 10. Splitting of the phase characteristic of PRC with clockwise rotation.

The results of the ring interferometer characteristics analysis show that the phase characteristic has distinctive features near the natural frequencies of the resonator [11]. It is impossible to directly measure the phase of the optical signal, however, phase information can be obtained as, for example, in fiber-optic gyroscopes [3] from the interference pattern. It is clear that there are two possible applications of the phase response of the PRC in MOG. First, you can use only the phase response to determine the difference in the natural frequencies of the PRC (to measure the angular velocity). This option can be implemented as follows: use a PRC with a high (close to 1) optical coupling between the resonator circuit and the auxiliary waveguide and, at the same time, either send signals from two PRC outputs to one photodetector, or send to each photodetector a signal from one of the PRC outputs and reference signal sent beside PRC. With a high value of the optical coupling coefficient the amplitude characteristic of PRC becomes almost constant (practically independent of the emission frequency) and does not affect the measurement result. In this case, MOG scheme actually degenerates into the Sagnac interferometer and therefore, from a practical point of view, the first version is not of particular interest. Secondly, it is possible to use both PRC phase and amplitude characteristics to determine the difference between the natural frequencies of PRC. As will be shown later, the simultaneous use of the amplitude and phase characteristics of PRC makes it possible to achieve some advantages over using only the amplitude characteristic.

Consider the method realization for measuring the angular velocity using the phase and amplitude characteristics of PRC [22]. The operating principle of such a resonant MOG resembles the principle of operation of the MOG considered above using only the amplitude characteristic (see Figure 4) and is explained in Figure 11. The radiation from laser 1 is divided by splitter 2 into two equal-intensity waves W1 and W2. The frequency of the radiation of both waves changes with the passage of the systems of modulators 3 and 4, which allows to scan the PRC spectrum. Phase modulators are controlled by the computing system 13. Then, the radiation in each of the optical channels is again divided into two equal parts by splitters 9 and 10. Wave W1 is divided into W11 and W12, and wave W2 is divided into W21 and W22. Waves W12 and W22 are not sent to the PRC and are used as reference waves. Waves W11 and W21 are directed to a directional coupler 11 connecting PRC 12 to the MOG optical channels. With its passage, the amplitude and phase of the waves changes. Then waves W22 and W21 are sent to splitter 9, and waves W12 and W11 to splitter 10, where they interfere. Then, with the help of directional couplers 7 and 8, the light is directed to photoreceivers 5 and 6, which acquire, respectively, the interference of waves W22 and W21 and waves W12 and W11. Thus, in the considered MOG scheme, in contrast to the scheme assuming the use of only the amplitude characteristic (Figure 4), the PRC is connected through a Mach-Zander interferometer formed by two splitters.

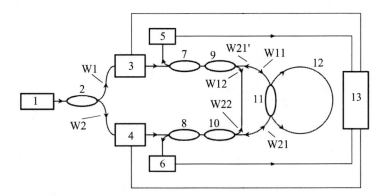

Figure 11. The operating principle of the resonator micro-optical gyro using the passive ring resonator phase and amplitude characteristics.

Let us analyze the signals arriving at photodetectors 5 and 6. For this, it is sufficient to consider the change in the optical signal during the passage of one of the channels of the circuit shown in Figure 11.

Let the electric field strength of wave W2 to be equal to E_{W2}, then the strengths of waves W21 and W22 are equal to:

$$E_{W21} = |E_{W21}| \cdot \exp(-j\beta l_{21}),$$
$$E_{W22} = |E_{W22}| \cdot \exp(-j\beta l_{22}),$$

where $|E_{W21}| = |E_{W22}|$ are amplitudes of waves W21 and W22 intensities, respectively; β is the propagation constant of light in the medium (we assume that the waves propagate through waveguides with equal propagation constants); l_{21} and l_{22} are the paths traveled, respectively, by waves W21 and W22 in the arms of the Mach-Zander interferometer. With the passage of the PRC, the electric field strength changes. Denote the PRC output wave W21'. Then its intensity is [22]:

$$E_{W21'} = E_{W21} \frac{\sqrt{1-K_c} - \sqrt{1-P}\exp(-j\varphi_{CW})}{1-\sqrt{1-K_c}\sqrt{1-P}\exp(-j\varphi_{CW})} = |E_{W21'}|\exp\left(-j\cdot[\beta l_1 + \Phi_{CW}]\right),$$

where K_c is the energy coefficient of the optical coupling between the cavity circuit and the auxiliary waveguide; P is the fraction of power lost in a single PRC bypass (energy loss); φ_{CW} is the phase advance with a single PRC clockwise bypass; $|E_{W21'}| = \sqrt{T_{CW}}\cdot|E_{W21}|$ is the amplitude of the intensity of wave W_{21}'; T_{CW} is the energy transfer coefficient of the CW wave through the section with PRC (PRC amplitude response when it is passed clockwise); Φ_{CW} is the PRC phase response for a clockwise bypass. Then two coherent monochromatic waves W22 and W21' are brought together and interfere. It is obvious that the radiation intensity at photodetector 5 is described by the expression:

$$I_5 = |E_{W21'}|^2 + |E_{W22}|^2 + 2\cdot|E_{W21'}|\cdot|E_{W22}|\cdot\cos(\beta\Delta l + \Phi_{CW}), \quad (7)$$

where Δl is the difference between the lengths of the paths of reference wave W22 and wave W21' (the difference of the lengths of the Mach-Zander interferometer arms). Earlier it was noted that the energy of wave W2 is divided into W22 and W21 equally and, therefore, $|E_{W21}| = |E_{W22}| = 0.5|E_{W2}|$. Given this fact, expression (7) can be converted as follows:

$$I_5 = 0.25\cdot(1+T_{CW})\cdot|E_{W2}|^2 + 0.5\cdot T_{CW}\cdot|E_{W2}|^2\cdot\cos(\beta\Delta l + \Phi_{CW}). \,(8)$$

Similarly, the expression for the radiation intensity on photodetector 6 can be modified:

$$I_6 = 0.25 \cdot (1 + T_{\text{CCW}}) \cdot |E_{\text{W1}}|^2 + 0.5 \cdot T_{\text{CCW}} \cdot |E_{\text{W1}}|^2 \cdot \cos(\beta \Delta l + \Phi_{\text{CCW}}) \quad (9)$$

where E_{W1} – electric field strength of W1 wave; T_{CCW} is the energy transfer coefficient of a CCW wave travelling through a section with a PRC (PRC amplitude response when it is passed counterclockwise); Φ_{CCW} is the PRC phase response when it is counterclockwise.

When the optical difference between the lengths of reference and measuring arms of the Mach-Zander interferometer is proportional or equal to an integer number of wavelengths λ_m, then the dependence of the intensity of optical radiation directed to photodetectors 5 and 6 on its frequency (output characteristic of the MOG optical path) corresponds to line 1 in Figure 12. The eigenfrequencies of the resonator correspond to the minima of the intensity of optical radiation. By the minima of the signals from photodetectors 5 and 6, the computing system 13 determines the PRC eigenfrequencies for mutually opposite directions of its bypass, their difference and, finally, determines the angular velocity. It should be noted that if the optical difference between the lengths of the reference and measuring arms of the Mach-Zander interferometer is not equal or not proportional to an integer number of wavelengths λ_m, then the shape of the output characteristic of the MOG optical path changes (see Figure 12). To facilitate the perception of Figure 12, the graphs corresponding to the difference between the lengths of reference and measuring arms from the interval from $\lambda_m \cdot (N + 1/2)$ to $\lambda_m \cdot (N + 1)$ are not given, since they are similar (symmetrical about the natural frequency of PRC) to the graphs from the interval from $\lambda_m \cdot N$ to $\lambda_m \cdot (N + 1/2)$. However, for any difference in the lengths of the arms of the Mach-Zander interferometer, it is possible to determine the natural frequencies of PRC from characteristic changes in the intensity of radiation incident on the photodetectors. In the resonator MOG scheme utilizing the phase response of a PRC, that was discussed above, one optical communication loop (one auxiliary waveguide associated with the resonator) is used to input and output radiation from the PRC. However, as for MOGs using only the PRC amplitude response of others configurations are possible (see Figure 7) that differ in the scheme of input and output of optical radiation from the resonator.

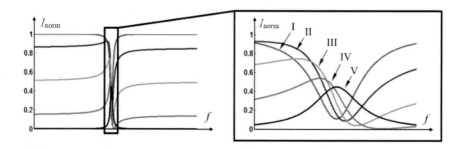

Figure 12. Normalized radiation intensity at photodetectors 5 and 6 with the difference of reference and measuring arms of the Mach-Zander interferometer equal to: $\lambda_m \cdot N$ – line I, $\lambda_m \cdot (N+1/8)$ – line II, $\lambda_m \cdot (N+1/4)$ – line III, $\lambda_m \cdot (N+3/8)$ – line IV, $\lambda_m \cdot (N+1/2)$ – line V.

Note that using the PRC phase response in addition to the amplitude one to determine the difference in the natural frequencies of PRC also complicates the PRC connection circuit. However, the described MOG scheme has advantages. It is known that the depth (when working on reflection) of the dips and the steepness of the PRC amplitude characteristic is determined by the optical coupling coefficient K_c and the losses in PRC. As noted above, while reducing the level of losses in a PRC, its Q-factor increases and the potential sensitivity of a MOG, based on it, increases. But it is obvious that with a fixed value of the coupling coefficient and a decrease of the level of loss, the depth of the dips will decrease and at $P = 0\%$ (in the absence of losses in the PRC or full compensation) the amplitude characteristic will be the same at all frequencies, and it will not be possible to determine its own PRC frequencies.

The PRC phase characteristic has distinctive features at its own frequencies even with full compensation of losses [11]. Considering this, it is easy to see from expressions (8) and (9) that, in contrast to a resonator optical gyroscope using only the amplitude response of a PRC, the circuit shown in Figure 11 allows to measure the angular velocity even with full compensation of the PRC losses. Also, the combined use of the phase and amplitude characteristics of PRC sometimes allows to improve the accuracy of the angular velocity measurement and the more significantly it is the smaller the losses in the PRC are [11].

Interferometric MOG

The operating principle of the interferometric MOG is similar to a conventional FOG, but the Sagnac interferometer used as a sensitive element is formed not by a fiber coil, but a spiral waveguide coil (Figure 13) or a PRC operating in a dual-beam interferometer mode (Figure 14). In an interferometric MOG, the radiation is driven into a two-beam interferometer in mutually opposite directions (CW and CCW), bypasses it and interferes at the output. Thus, a photodetector located at the output of the interferometer records the result of the two-beam interference of the waves that bypassed the interferometer along the CW and CCW directions:

$$I = 2I_0(1 + \cos \Delta\varphi), \tag{10}$$

where I_0 is the intensity of light bypassing the dual-beam interferometer in one of the opposite directions; $\Delta\varphi$ is the phase difference of the waves bypassing the dual-beam interferometer along the CW and CCW directions. At the same time, the difference between the phases of the counterpropagating waves $\Delta\varphi$ caused by the Sagnac effect is determined by expression (5) and is proportional to the angular velocity. Thus, a change in the angular velocity leads to a change in the intensity of the radiation on the photodetector. According to the indications of the photodetector, the Sagnac phase difference $\Delta\varphi$ and the angular velocity Ω are judged. However, as follows from expression (10), at low angular velocities, the intensity on the photodetector varies slightly with changing speed, i.e., the sensitivity to low angular velocities is low. In order to increase the sensitivity at low speeds and to ensure linearity of the output characteristics in the interference MOG, methods once developed for FOG are used: mutual phase modulation and a closed signal processing circuit [3, 23]. To implement this, a phase modulator is introduced into the dual-beam interferometer, with which the phase shift between the counterpropagating waves from $\pi/2$ to $-\pi/2$ with a frequency of $1/2\tau_0$ is modulated, where τ_0 is the time of the fiber circuit light bypass (for mutual phase modulation). Also, with the help of this modulator, an additional phase shift between counterpropagating waves equal in magnitude to the Sagnac phase shift φ_S and opposite in sign (for a closed signal processing circuit) is created.

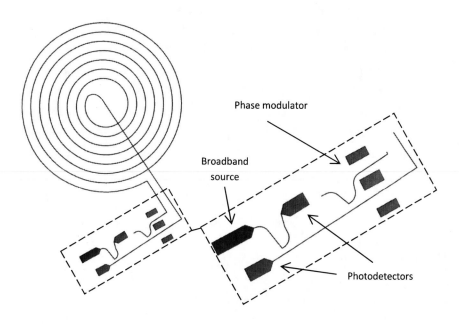

Figure 13. Interferometric MOG with a spiral waveguide coil.

Usually interferometric MOG are implemented using a helical waveguide coil. The ends of the waveguide coil approximate to each other at a distance of the order of microns and are arranged along the same direction. They form a directional optical coupler with a branching factor of 50%. Thus, a coil with a coupler form the Sagnac interferometer. The mutual configuration of such a MOG is schematically depicted in Figure

13. In the interferometric MOG, a broadband light source is used to minimize the effect of backscatter in a spiral waveguide on the measurement results. One of the photodetectors is located at the input of the interferometer (opposite the light source) and is used to control the input radiation power. The second photodetector is used to measure the signal from the output of the interferometer. Phase modulators are used to create mutual phase modulation of the oncoming waves and implement a closed signal processing circuit. The described configuration of the interferometric MOG was recently studied in article [24]. In this case, a spiral-shaped resonator with a length of 10 m, a helix pitch of 50 μm and a minimum bending radius of 1 mm was analyzed. The potential sensitivity of such a gyroscope (with an area of about 10 cm^2) was estimated at about 19 °/h with a loss in the waveguide of 1 dB/m. In the studies of other research groups devoted to this type of interferometric MOG, exactly the same switching circuit of the sensitive element is used [25-27]. In this case, the differences is mostly in the used components. It is worth noting that a serious design disadvantage of the interferometric MOG utilizing a waveguide coil is the intersection of the turns of the waveguide spiral by one of its ends, which leads to the appearance of additional losses in the interferometer [24].

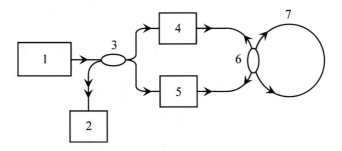

Figure 14. Re-entrant MOG.

There are also Interferometric MOG configurations, in which a ring waveguide (PRC) is used instead of a waveguide spiral coil [28-31]. However, in this case, the ring waveguide is not used as a resonator, but as a delay line. This configuration of the interference MOG is schematically shown in Figure 14. In MOG the result of the interference of two waves that bypass the closed loop of the PRC (7) in mutually opposite directions is recorded. The beam from the light source (1) is divided by a directional coupler (3) into two parts and passes through a system of modulators (3) and (4), which create a controlled phase shift between the opposing waves. Radiation is introduced into the PRC (usually a waveguide ring) through the directional coupler (6) in the clockwise and counterclockwise directions. Waves that passed the PRC are derived from it and sent to the photodetector (2), which registers their two-beam interference. The PRC in this case plays the role of a multi-turn coil. Its equivalent number of turns N corresponds to the number of rounds of the PRC by light wave. The number of rounds is determined by the loss in the waveguide and its perimeter. In order for light streams coming after different number of rounds to not interact with each other (to avoid multipath interference), pulse phase modulation is used. The pulse width is chosen equal to the circuit light bypass circuit, and the repetition frequency period is N + 1 times more [31]. It is worth noting that this configuration is less sensitive to changes in angular velocity than the interferometric MOG with a spiral waveguide coil and the resonator MOG [28]. Interferometric MOG configurations utilizing a closed waveguide instead of a spiral waveguide coil are sometimes called interference-resonator, or return MOG.

CLASSIFICATION OF PASSIVE MICRO-OPTICAL GYROS ACCORDING TO THE STRUCTURE OF THE SENSITIVE ELEMENT

The sensing element is the basis of MOGs. They determine the manufacturing technology of the entire instrument, the potential sensitivity, the minimum possible dimensions and many other characteristics of the gyroscope. As noted above, spiral-shaped waveguide coils or PRCs can be used as sensitive elements of passive MOGs. Spiral waveguide coils are used less frequently in MOGs and differ from each other only in the number of turns and the materials used in their manufacture. As a rule,

PRCs are used as the MOG sensing element. Usually, when developing MOGs, waveguide PRCs are used as a sensitive element. Most often they are made of planar waveguides. They can also be made of fiber or photonic crystals. In this case, only single-mode waveguides are used. As in all optical gyroscopic systems, the use of multimode waveguides is impossible because of the mode dispersion. For the same reason, conventional open resonators cannot be used. Although it is possible to use one of the specific subspecies – confocal ring resonators. The resonators of the whispering gallery modes stand apart. They have recently been considered as promising sensitive MOG elements. Thus, the structure, properties and manufacturing techniques of PRC may differ significantly. Therefore, MOG is conveniently classified based on the type of PRC used (see Figure 15).

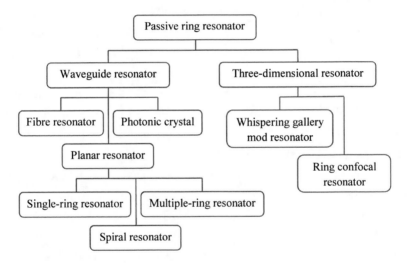

Figure 15. Types of passive ring resonators, considered as sensitive elements of micro-optical gyros.

Single Ring Planar Passive Ring Resonators

Nowadays integral (planar) optics technologies are mostly used in the development and manufacture of both individual elements and the entire MOG as a whole. The use of integrated-optical technologies in the device manufacture provides a number of advantages: 1) all MOG-elements can be compactly placed at the integrated circuit (all elements of the MOG can be placed even inside the PRC); 2) weight reduction (all elements may have a common frame if placed on one integrated circuit): 3) power consumption reduction; 4) mass MOG production can be organized; 5) increase of MOG reliability (the whole MOG can be manufactured as a single rigid structure – an integrated circuit), etc. [10]. Usually single-mode waveguide planar single-ring resonators are used as PRC in MOGs. Single-ring resonators are annular waveguides (closed waveguides usually having form of a ring or a track) optically linked to an auxiliary waveguide, which is used for input and output of radiation. They are relatively simple to be manufactured, have a quality sufficient for MOG, are reliable, and their manufacturing process is well established. It is known that the potential sensitivity of the MOG based on the PRC is proportional to the product of quality factor of PRC and its diameter [3, 23]. In order to achieve a potential sensitivity of 10 °/h or less when using PRC with a diameter of no more than a few centimeters, a PRC with quality factor surpassing 10^5 should be used. The current level of technological development makes it possible to produce suitable high-quality MOG PRC from silicon, quartz, silicon nitride, lithium niobate, indium phosphate and optical polymers.

Compact high-quality PRCs can be obtained in quartz waveguides grown on silicon substrates. A thin layer (several microns thick) of silicon oxide is applied onto the substrate. Then a thin (a few microns) layer of quartz, slightly doped by phosphorous, germanium oxide, etc., is spread by means of plasma-chemical deposition from the gas phase. After that a pattern made of protective coating is applied by means of ultraviolet lithography, then unprotected areas of the structure are removed up to the substrate by means of ion etching. Afterwards a protective silicon oxide layer is applied over the entire structure using plasma-chemical deposition from the gas phase. After applying each new layer, the structure is subjected to thin annealing. Using this technology PRCs with a ring radius of 30 mm and losses less than 0.85 dB/m (corresponding to Q-factor of 2.3×10^7) were constructed [32]. The potential sensitivity of a MOG using such quartz PRC is 1.6 °/h [33]. Also, according to this technology, several more compact resonators were constructed applying this technology: a resonator with a ring radius of 17.5 mm and a Q-factor of more than 6 × 10^6 [12]; a resonator with a radius of 12.5 mm and a Q-factor of the order of 8 × 10^6 [34]. It should be noted that structures manufactured in accordance with this technology are characterized by low losses (high Q-factor) only if PRC diameters are of the order of a few centimeters. If diameter is less than 1 cm, the quality factor drops significantly due to an increase of the waveguide curvature caused by small difference between the refractive indices of the waveguide core and the cladding, so the radiation cannot retain in the core.

High-quality PRCs can also be made of silicon nitride. A silicon dioxide layer of 10–15 μm thickness is deposited on quartz or silicon substrate by magnetron sputtering. Then a submicron (from tens to hundreds nm) layer of silicon nitride is deposited by means of chemical deposition from the gas phase under reduced pressure. After this, photolithography is applied to form a protective pattern and using ion etching or plasma etching an unprotected part of the submicron layer is removed. Finally, the structure is covered with a protective silicon dioxide layer of 10–15 μm thickness [35]. PRCs with a ring radius of 9.8 mm and losses, less than 0.6 dB/m (which corresponds to a quality factor of 5.5 × 10^7), were constructed using this technology [36]. Later the result was improved: for a radius of 9.7 mm, the quality factor was 8.1 × 10^7 [37]. Moreover, it was experimentally demonstrated that a loss of 0.1 dB/m can be achieved in silicon nitride waveguides with radii of curvature up to 7 mm [37].

More compact high-quality PRCs can be obtained in silicon waveguides on an optical insulator (SIO structure). Such waveguides are made from blanks consisting of a quartz substrate and a several microns thick silicon layer. Silicon layer thickness can be reduced to 1.33 microns (this thickness provides a single-mode regime) by realizing oxidation in water vapor, followed by oxide removal. A protective pattern is applied to the waveguide structure with the help of photolithography. Then silicon is etched until its thickness reaches 1.1 microns everywhere, except the protected part of the structure whose thickness remains 1.33 microns. As a result high-quality comb waveguides are obtained. To protect the structure from external impact, it is coated by a layer of silicon oxide. PRCs with ring radius of 2.45 mm and losses, less than 2.7 dB / m (corresponding to a Q-factor of 2.2×10^7), were constructed, applying this technology [38].

Lithium niobate-based PRCs have worse Q-factor if compared to PRCs, made of any of abovementioned materials, and having equal sizes. However, they are simpler and cheaper to be manufactured; moreover, it is easy to introduce additional impurities into the waveguide, such as atoms of active substances for compensation of PRC losses. Fabrication of lithium niobate waveguide structures starts with formation of titanium coating on a blank, the thickness of a film is several tens of nanometers. Then a protective pattern is formed on the waveguide structure by means of photolithography. Titanium layer that is not protected by photoresist is removed by ion etching. Then the sample is placed into a furnace for 10 hours, where diffusion of titanium into a lithium niobate substrate occurs at temperature of about 1020°. Applying this technology, PRCs with a ring radius of 30 mm and losses, less than 3.0 dB / m (corresponding to a Q-factor of 2.4×10^6), were constructed [39]. Comb waveguides can also be made of lithium niobate. PRCs with radii of about 100 microns and a quality factor of about 10^7 can be constructed, using such comb waveguides [40].

Indium phosphate-based waveguide structures can also be applied in construction of high-quality resonators. It is relatively easy – compared to quartz, silicon and silicon nitride structures – to integrate all of the optical and optoelectronic MOG components [41-43] on a single substrate when using this material. To fabricate high-quality PRCs, based on this material, firstly, a 1.3 microns thick InGaAsP layer (thickness should provide a single-mode mode) is grown on an indium phosphate substrate, using MOVPE technology. Next, a protective pattern is applied to the waveguide structure by means of lithography and then InGaAsP layer thickness is reduced to 1 μm everywhere except the protected part by reactive ion etching. According to this technology, a PRC with a radius of 13 mm and a Q-factor of 0.97×10^6 was fabricated. It was demonstrated that usage of such PRC makes it possible to obtain the potential sensitivity of MOG of 10 °/h [44]. Moreover, PRCs with loss compensation can be implemented using waveguides made of this material. Due to this, it is possible to achieve a quality factor of about 1.5×10^7 for a resonator radius of about 5 mm [20].

The simplest and cheapest PRCs can be obtained using optical polymers [45-47]. Losses in optical polymers are higher than in the previously discussed materials. However, relatively recently, a polymer PRC with a 10 mm diameter and a Q-factor of about 10^5 was developed [47]. The waveguide structure of the polymer PRC is fabricated on a silicon substrate. A 15 μm thick layer of optical polymer with a refractive index of 1.45 is deposited on a substrate; a waveguide made of another optical polymer with a cross section of 5×5 μm and a refractive index of 1.46 passes at a distance of 5 μm from the substrate. Recently, polymer resonators with radii of about 100 μm and a Q-factor of about 8×10^5 have been manufactured [48].

Multi-Ring Planar Passive Ring Resonators

Currently, micro-optical multiring resonators are considered as a sensitive element of MOG [49-64]. Two basic structures are distinguished. The first one (Figure 16a) is a chain of optically coupled resonators (CROW). The second structure (Figure 16b) consists of several ring resonators optically connected with one supplementary waveguide (SCSSOR).

As a rule, chain-shaped resonators consist of an array of high-quality miniature (diameter from up to tens of microns) PRCs. The adjacent array resonators are connected to each other by tunneling (due to frustration of total internal reflection), forming a single waveguide (the so-called CROW). It is proved that the sensitivity of these structures usually does not exceed the sensitivity of a MOG on a single-ring resonator, in case both resonators have equal areas [49-54]. Alongside, some researches suppose that in some cases usage of a CROW can result in achievement of greater sensitivity than sensitivity of single-ring PRCs [55-57]. New CROW designs, developed for sensitivity increase, are theoretically analyzed [57, 58]. However, these studies do not take into account the fact that the sensitivity of real (not idealized multi-ring resonators) devices is reduced due to the greater losses if compared to single-ring PRCs [54]. Specific losses of multi-ring resonators exceed the losses of single-ring resonators of equal area, since the size of the components of the CROW resonators is smaller, which results in higher curvature and radiative losses. In multi-ring PRCs, additional losses occur in regions where tunneling connection of resonators takes place. Nevertheless, multi-ring resonators are of interest as sensitive MOG elements, in case high-quality resonators of only small size are possible to be fabricated.

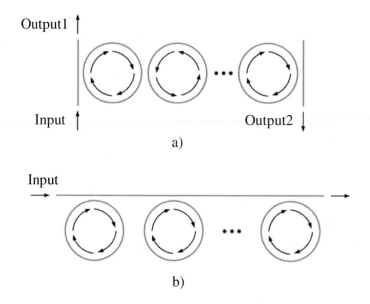

Figure 16. Multi-ring resonators: a – CROW, b – SCSSOR.

The design of developed CROWs is extremely diverse. Various configurations of gyroscopes based on chain-like resonators are known. Often, the simplest configuration, corresponding to Figure 16a [55, 56, 59], is discussed. Much attention is also paid to the structure consisting of a curved chain of identical PRCs (Figure 17). The coupling coefficients between the PRC are the same. Auxiliary waveguides, optically coupled to the end resonators, terminate in a directional coupler. The input signal is divided by the coupler into two waves, detouring the resonator in opposite directions. Returning to the coupler, the opposite waves interfere and, then, are directed to the output structure [57]. Thus, an interferometric scheme of the sensitive element connection is used.

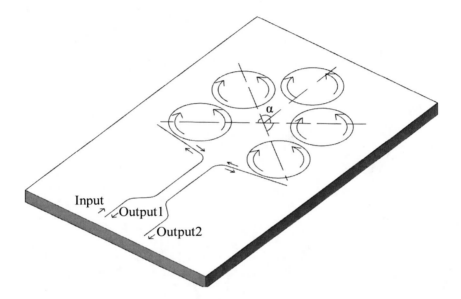

Figure 17. Curved chain-like resonator scheme.

The scheme considered in [60] is somewhat different: identical optically coupled ring resonators are arranged in a straight line. The coupling coefficients between the waveguides are different (take one of two values) and alternate, starting from the center, as shown in Figure 18 (dark rectangles indicate a stronger optical connection, light – indicate less strong optical connection). Theoretically, such distribution of coupling coefficients, if compared to a uniform one, makes it possible to increase the potential sensitivity of the device to the rotation by several orders of magnitude. At certain values of the coupling coefficients of strongly and weakly bound PRCs, it is possible to increase the sensitivity as compared to a CROW having identical coupling coefficients [60, 61].

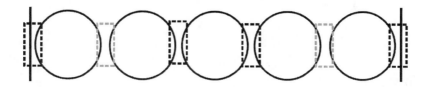

Figure 18. Chain-like resonator with alternating coupling coefficient values.

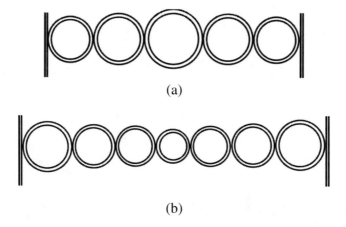

(a)

(b)

Figure 19. CROWs formed by ring resonators, that differs by an integer number of wavelengths (a) resonator perimeter decreases uniformly from a center to a circumference, (b) resonator perimeter increases uniformly from a center to a circumference.

From the point of view of optimizing the sensitivity to the rotation, various configurations of chain-like resonators composed of PRCs of different sizes are of interest (see Figure 19a, b) [58, 62]. In this case configurations consisting of an odd number of ring resonators (ring waveguides) located symmetrically with respect to the center of the chain-like structure are considered. Attention is focused on chain-like structures, in which the perimeters of any two adjacent ring resonators differ by an integer number of wavelengths [58]. It was analytically shown that such configurations of chain-like resonators allow to increase the sensitivity to the rotation if compared to a standard configuration (Figure 19a). For example, the potential sensitivity of a gyroscope with a chain-like resonator, composed of five circular waveguides, whose perimeter differs by one wavelength, corresponds to the potential sensitivity of a gyroscope with a chain-like resonator of a standard structure, composed of 35 rings [58].

It should be mentioned that a number of problems associated with practical implementation of gyroscopes based on chain-like resonators exist. Random deviations of the ring resonators sizes and the coupling coefficients occur when these structures are being manufactured. However, random deviations of coupling coefficients in chain-like structures practically do not affect the sensitivity to the rotation [56]. However, even small fluctuations of the ring resonators sizes significantly reduce the sensitivity of the device [56, 58, 62].

Another type of multiring resonators (SCSSOR) is not so sensitive to the fluctuations of sizes of PRCs it consist of. Sizes oscillations of the ring resonators, SCSSOR consist of, within 0.1 μm do not practically affect the gyroscope sensitivity [63]. However, the SCSSOR sensitivity is lower than sensitivity of an equivalent single-ring PRC [49], as it was in the case of a CROW. Let us consider various types of moniliform resonators.

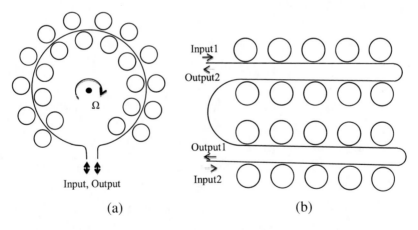

Figure 20. Moniliform resonators: (a) ring resonators are located on both sides of the auxiliary waveguide, (b) ring resonators are located on one side of the auxiliary waveguide.

The first type of SCSSOR resonators assumes that each of the oncoming waves passing through it would bypass some of the components of the SCSSOR ring resonators in a clockwise direction, and some – in counterclockwise. This configuration was discussed in [64]. The SCSSOR structure included an auxiliary waveguide, bent in the form of a ring, and numerous high-quality micro-optical ring resonators, located on both sides of the auxiliary waveguide (Figure 20a). Both ends of the auxiliary waveguide were used for the input and output of signal. Authors proposed to use WGM (whispering gallery modes) resonators as components of the high-quality resonators structure [64].

Another type of moniliform resonators assumes that each of the oncoming waves passing through it would bypass all constituting ring resonators in one direction. Configuration of such moniliform resonator was discussed in [63]. Unlike the previous one, in this configuration all ring resonators are connected with only one side of the auxiliary waveguide (Figure 20b). It was proposed to use this structure in the interference scheme of the sensitive element inclusion.

Spiral Sensitive Elements

Another promising type of MOG sensing element is a spiral waveguide resonator. The idea of using a spiral waveguide coil for creation of a cheap and compact gyroscope has been existing for a long period of time. However, this type of resonators has appeared in the focus of interest relatively recently. It can be explained by the difficulty of manufacturing of a miniature spiral waveguide coil (curvature radius of several mm or less), characterized by low propagation losses; it requires application of modern integrated optical technologies [65]. For this purpose spiral sensitive elements of two types are used as MOG sensitive elements: passive ring resonators and dual-beam interferometers [24, 65].

The first one (see Figure 13) is a two-beam interferometer [24, 66]. Spiral waveguide's ends are spaced closely (order of one wavelength), forming a directional coupler due to the optical tunneling effect. The coupling coefficient between the ends should be close to 50%, otherwise the structure's sensitivity to the rotation decreases drastically [24]. It was previously mentioned that MOGs based on such a waveguide act as a waveguide version of a FOG and refers to interference type. Existing level of development of integrated optics makes it possible to fabricate interference MOGs based on a waveguide coil occupying on a substrate less than 10 cm², possessing a potential sensitivity of 19 °/h [24]. Notice, that the area occupied by the coil can be significantly reduced by connecting several spiral waveguides in series and placing them in a column on each other. In this case, the layers can be joined with the help of vertically arranged directional couplers. However, in practice the resonator losses increase due to the losses arising from the transition of light between adjacent layers (0.02 dB per one transition), which results in device sensitivity reduction [24]. Recently, a compact configuration of such sensor based on silicon waveguides with dimensions of 600 μm × 700 μm was realized. The potential sensitivity of the gyroscope based on it is 51.3 °/s (approximately equivalent to 18×10^4 °/h) [66]. In perspective, the losses on the transitions between layers can be significantly reduced due to the development of integrated-optical technologies and the use of multilayer topology may become worth.

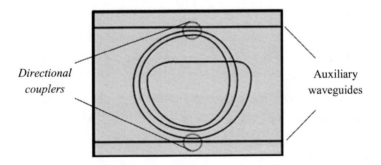

Figure 21. Sensor based on a spiral PRC.

Another configuration of a spiral waveguide sensor is a passive ring resonator (see. Figure 21). Recently results of experimental study of a MOG based on a spiral PRC has been published [65]. A circuit consisting of a PRC (spiral closed-loop waveguide) and two auxiliary waveguides, connected with it by directional couplers, was investigated. The spiral was made of a single-mode (6 μm × 6 μm cross-section) silicon waveguide doped by germanium (refractive index 1.457). The length of the spiral waveguide was 42 cm; area, occupied on a substrate, was 20 cm^2. The propagation losses in a spiral waveguide were 10 dB/m. Additional losses were observed; they are associated with the intersection of the helix (less than 0.01 dB). The Q-factor of the described resonator was experimentally defined, it was about 1.5×10^6. The potential sensitivity of a MOG, based on a PRC, is 156 °/h [65]. Later a more compact high-quality spiral PRC, based on indium phosphate, was fabricated [67]. Its Q-factor was about 6×10^5 while area, occupied on a substrate, was 10 mm^2 (diameter ≈ 3 mm). The potential sensitivity of MOG based on it was estimated to be 10 °/h [67].

Fiber Passive Ring Cavities

Not long ago, a compact high-quality fiber PRC was developed, manufactured and studied [68]. It is a miniature version of a fiber resonator, which was developed in 1982 and consisted of a single turn of single-mode fiber connected to the directional coupler (Figure 22a) [69]. The difference from the miniature version (Figure 22b) is that the role of the directional coupler is performed by the region of the resonator in which the fiber sections are overlapping. Since the loop is folded on the fiber section where the shell thickness is preliminarily reduced (either by etching or pulling), when fiber sections are overlapping of the total internal reflection effect violation occurs and an optical coupling loop is formed (directional coupler).

To date, micro-optical fiber resonators were obtained with a radius of about 0.5 mm and a quality factor of more than 1×10^6 with a fiber cross-section diameter (in the fiber loop area) around 1 μm [68, 70]. The production method of such resonators consists of pulling and twisting of the microfiber into an optically connected loop. When the microfiber is pulled, it is placed inside a sapphire capillary that is heated by a beam of an industrial laser (for example, a CO_2 laser). The sapphire capillary plays a role of a microscopic furnace, inside which the fiber is drawn, until its diameter reaches the required size (around 1 µm or less) (Figure 23a). Then, with the help of several manipulators, the obtained biconical fiber is twisted into an optically connected loop (Figure 23b). Surface forces (van der Waals and electrostatic) [68] help to connect the loop ends.

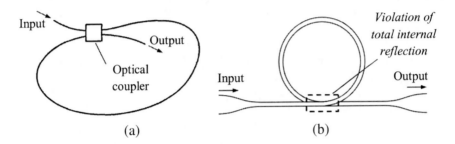

(a)   (b)

Figure 22. Fiber PRC. (a) Macroscopic design, (b) Microscopic design.

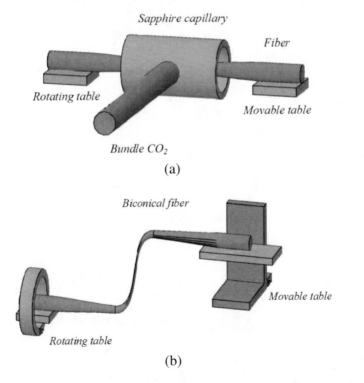

Figure 23. Production of a micro-optical looped fiber PRC. (a) Stretching fiber, (b) twisting of a loop-like resonator.

After manufacturing, such a resonator can be coated with a polymer with a low refractive index for protection against mechanical stress [71].

It is worth noting that, compared to the waveguide PRCs, fiber loop-shaped resonators still have the maximum achievable quality factor 1-2 orders of magnitude lower then waveguide resonators. In practice, the accuracy of the resonator MOG is limited by the errors caused by the backscattering effect of the MOG waveguide structure and fluctuations of the opposing waves polarization (polarization noise). The noise caused by the backscattering can be reduced below the potential sensitivity level using double phase modulation technique [13]. Planar waveguides used to create sensitive elements are generally characterized by small losses for only one polarization state. In the discussed above fiber PRCs waves with two orthogonal polarization states can propagate simultaneously. Therefore, when using them as sensitive elements of resonant MOGs, due to the high level of polarization noise, in practice the achieved accuracy will be much lower than the one achieved using similar planar PRCs. Polarization noise in MOGs on fiber resonators can be reduced by using polarization-maintaining optical fiber (PM fiber) [72]. The use of resonators made of PM fiber makes it possible to achieve good results [73, 74], but resonators made of such fiber cannot be miniaturized yet (manufactured ones are with diameters more than 10 cm).

Photonic Crystal Passive Ring Resonators

Photonic crystals are materials with ordered structure characterized by a strictly periodic variation of the refractive index with its period comparable to the wavelengths in the visible and near-infrared ranges [75]. In this case, as shown in Figure 24, photonic crystals are distinguished with a periodic change of the refractive index of the medium in one, two, or three dimensions (one-dimensional, two-dimensional, and three-dimensional photonic crystal structures, respectively).

The photonic crystals, due to the periodic refractive index variation, make possible to obtain the allowed and forbidden zones for photon energies, similarly to semiconductor materials, in which allowed and forbidden zones are observed for the energies of charge carriers. Practically, this means that if a photon with energy (frequency) that corresponds to the forbidden zone of a given photonic crystal falls on a said crystal, it cannot pass through the crystal and is reflected back. And vice versa, if a photon with an energy that corresponds to the allowed zone of a given crystal falls on a surface of this crystal, it can propagate. Any inhomogeneity in a photonic crystal structure is called a photonic crystal defect. In such areas the electromagnetic field is concentrated, and this is used in micro resonators and waveguides constructed using photonic crystals.

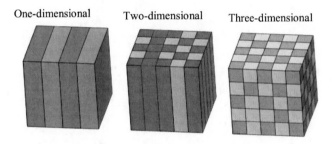

Figure 24. Photonic crystals types.

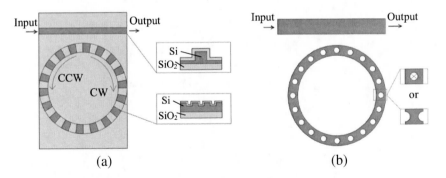

Figure 25. One-dimensional photonic crystal PRC. (a) ring-shaped Bragg diffraction grating, (b) closed-loop perforated waveguide PRC.

High Q-factor photonic crystal PRCs can be made from various types of one-dimensional photonic crystals [76, 77]. For example, in [76] a PRC is made using a ring-shaped Bragg diffraction grating (Figure 25a). Another type of a one-dimensional photonic crystal for the PRC MOG is a closed planar semiconductor waveguide, perforated with a one-dimensional periodic system of holes (Figure 25b) [77]. Analytically and with the help of modeling it was shown that at the present level of development of integrated optics it is possible to manufacture PRCs based on a one-dimensional photonic crystals with a radius of 2-10 mm and quality factor of more than 10^9 [76, 78].

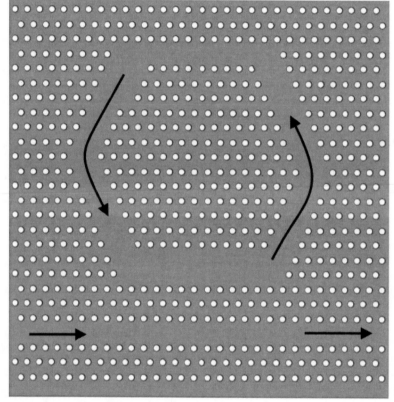

Figure 26. Two-dimensional photonic crystal PRC.

In addition, for the manufacturing of a high-quality (> 10^6) microoptical PRC a photonic crystal structure, made of a planar semiconductor waveguide, perforated by a periodic two-dimensional hole system, is very promising. Using such a structure, it is possible to create a photonic crystal PRC by introducing equilateral hexagon shaped defects into the structure (Figure 26). The thickness and size of the hexagons determine the composition of the eigenmodes and the quality factor of the resonator [79].

A few years ago, a study with demonstration of the Sagnac effect in photonic crystal resonators was published, where it was analytically shown that the rotation of the photonic crystal resonator causes the splitting of the resonant frequencies of the counterpropagating waves. The magnitude of the effect was also evaluated with specific examples of photonic crystals [80]. However, the sensitivity limit of such gyroscope, based on a photonic crystal resonator considered by the authors, turned out to be too small for practical demonstration of the effect. This is due to the fact that a photonic crystal PRC with a radius of about 0.93 μm and a quality factor of about 10^4 was considered [80]. Similar results were obtained while attempting to create a passive MOG based on a multi-ring two-dimensional photonic-crystal PRC [81]. The multi-ring configuration was used to increase the PRC area, which was 279 μm^2.

To date, photonic crystal PRCs were developed with a quality factor of more than 10^9 [76, 77]. The size of the photonic crystal PRC can be also increased. However, the technology of manufacturing of the waveguide photonic crystal structures is still poorly developed and photonic crystal PRCs can probably be considered as sensitive elements of MOGs only in the future.

Ring Confocal Resonator

In most experimental setups and MOG prototypes, waveguide passive ring resonators are used as sensitive elements. In this case, only single-mode waveguides are used, since the use of multimode waveguides in this case is impossible due to the mode dispersion. For the same reason most of open ring resonators cannot be used. However, there are confocal configurations of ring resonators with a strongly degenerate and equidistant spectrum [82]. The use of such configurations as sensitive elements makes it possible to avoid the negative effect of the mode dispersion of an open ring resonator.

In the first approximation, it is possible to obtain a confocal ring resonator using several (at least three) reflecting surfaces: flat and concave toroidal (at least one) with certain (ensuring the degeneracy of the spectrum) radii of curvature [82]. In this case, the number of used toroid-shaped surfaces will determine the number of beam waists in the resonator and, therefore, the ratio of the distance between adjacent transverse modes to the distance between adjacent fundamental longitudinal modes.

In the practical implementation of confocal ring resonator MOGs, the resonator can be made using a different elemental base. Either mirrors or prisms of total internal reflection can be used as reflecting surfaces (see Figure 27). In this case, the loss per revolution of open ring resonators can be reduced to less than 0.01% [83], which makes it possible to achieve a quality level of about 10^9.

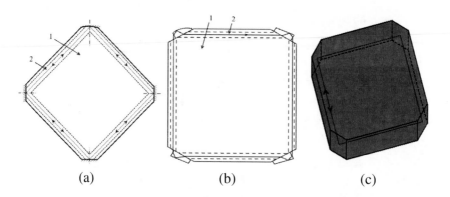

(a)   (b)   (c)

Figure 27. (a) Ring resonator composed of several mirrors, (b) of several reflective prisms and (c) of one reflective prism.

When fabricated using mirrors (Figure 27a), the reflecting surfaces are fixed on a rigid frame made of a thermally stable material 1 (Sitall, cerodur, invar, etc.). One of the mirrors used for the light input and output is semi-transparent. The light between the mirrors propagates either in vacuum or in the gaseous environment of capillaries 2. Thus, it is possible to get rid of major medium nonlinearities, its temperature fluctuations and other intracavity effects. However, when using a ring confocal resonator as a sensitive MOG element, the prism variant is preferable (Figure 27b, c). In this case, special reflective coatings are not used on the reflecting elements of the resonator. Therefore, in a prism version, the ring confocal resonator is distinguished by greater stability of operating parameters and greater reliability. In addition, the total internal reflection prisms are usually made of materials resistant to electromagnetic and radiation effects. The advantages of the prism design of the ring resonator include low sensitivity to acoustic and mechanical effects (which is difficult when using mirrors). Thus, the use of total internal reflection prisms instead of mirrors allows to obtain a resonator suitable for mass production. It should be noted that the prism version has one serious drawback. As a rule, three or four total internal reflection prisms are used in ring prism resonators (see Figure 27b). When the beam passes through each prism edge, a significant radiation loss is observed at the interface between the media (not all light passes – through the prism, some is scattered on the edge, some is reflected back). Therefore, to reduce the losses, the geometry of prism resonators is usually chosen in such a manner that the angle of incidence on the refracting faces of the prisms is equal to the Brewster angle [83]. This restriction in the choice of the resonator geometry in addition to the conditions of confocality of the ring resonator complicates the calculation of its design by an order of magnitude and makes the resonator more sensitive to re-adjustments due to changes in temperature and other external conditions. To eliminate this disadvantage a fully monoblock design can be used. When using a ring confocal resonator as a sensitive element of a miniature optical gyroscope, it is possible to manufacture a resonator from a single prism (Figure 27c). In this case, different faces of the same reflective prism are used as reflecting surfaces. The light input and output from the monoblock can be accomplished by violating the total internal reflection on one of the flat faces of the prism. It is worth noting that for some types of miniature optical gyroscopes (for example, those operating on a phase response of a passive ring resonator), replacing a waveguide ring resonator with a confocal one can provide some advantages [22]. The disadvantage of a ring confocal resonator is a relatively complex technology of making its concave astigmatic reflective surfaces.

Whispering Gallery Mode Resonators

The phenomenon of the whispering gallery has been known for a long time. It is based on the fact that sound (whisper) uttered in the premises sometimes spreads not along the shortest path, but along concave surfaces (walls, domes, etc.). As a result, it seems that the concave gallery surface "whispers". The term "whispering gallery mode" (WGM) was first used by Rayleigh in the 19th century to describe the phenomenon of whispering galleries under the dome of the cathedral in London [84]. At the beginning of the 20th century it was proved that electromagnetic WGMs exist. Recently, WGM optical resonators have attracted the continuously growing interest of the scientific community. They are dielectric axially symmetric resonators with smooth edges that support the existence of a WGM due to the total internal reflection on the surface of the resonator. To date, various types of such resonators have been developed: spherical, disk-shaped, toroid-shaped, bottle-shaped, etc. (see Figure 28) [85]. Increased interest is associated with their unique properties: ultra-high quality factor (more than 10^9), a limited number of eigenfrequencies, small dimensions (diameter from a few centimeters to tens of microns). These properties make it possible to use them as sensitive elements for compact high-precision devices, in particular, MOGs [86].

Recently, an experimental study of a MOG on WGM resonators [87, 88] was published. First, the MOG prototype was demonstrated and tested. The prototype used a disc-shaped crystal resonator made of fluorite with a quality factor of about 10^9, a disk diameter of about 1 cm and a thickness of 0.2 mm [87]. The radiation was introduced into the resonator at a wavelength of 1.5 μm with an intensity of 0.15 mW (0.075 mW was introduced into the resonator in each of the opposite directions). At the same time, in order to reduce the influence of external factors, the WGM resonator was isolated from the environment by a sealed enclosure. Measurements have shown that the ultimate accuracy of such MOG is 2.3 °/h. Then an experimental sample of this sensor was manufactured and investigated [88]. It used a similar WGM resonator with a diameter of 7 mm. The volume of the entire experimental sample, including the resonator, semiconductor laser, photodiodes, etc., was 15 cm³. Its maximum accuracy was about 3 °/h.

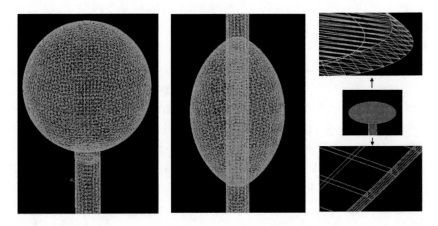

Figure 28. WGM optical resonators: spherical (left), bottle-shaped (center), disk-shaped (top right) and toroidal (bottom right).

The use of WGM resonators provides some advantages. It is known that the greater values of ring resonator Q-factor are, the higher angular velocity sensitivity limit of the sensor based on it is [3]. To date, the Q-factor achieved in the WGM resonator far exceeds the Q-factor of the waveguide resonator of equivalent size. The maximum Q-factor obtained in the WGM resonator from calcium fluoride was more than 10^{10} [89]. Typical Q-values for WGM resonators are of the order of 10^9. In addition, the use of WGM resonators instead of waveguide resonators reduces the error caused by backscattering, polarization noise and the Faraday effect [88]. Despite these advantages, there are serious drawbacks of the WGM resonators. In contrast to single-mode waveguide resonators, the spectrum of a conventional WGM resonator is weakly degenerate and, as a rule, has several modes with frequencies close to the used mode. The interaction between these modes is a source of errors [87]. It is known that the higher the input radiation power is, the higher sensitivity limit of the angular velocity sensor operating on the Sagnac effect is [3]. However, as a result of several factors (e.g., extremely high quality factor of the WGM resonators, the properties of the materials they are made of, and the fact that the radiation in them is localized in a very limited volume), an increase in power during the use of WGM resonators leads to the appearance of various non-linear effects. They lead to the appearance of an additional component of the error. This imposes a restriction on the input radiation power and makes it difficult to reduce the size of the angular velocity sensors with described WGM resonators to a diameter of the order of several mm or less [88]. In addition, the shape of the WGM resonators makes integration with other MOG elements (radiation source, photodetectors, etc.) difficult. Moreover, WGM resonators are sensitive to various external influences, ex., micromechanical gyroscopes [90, 91]. It should be noted that in addition to MOGs WGM resonators can be used as sensitive elements of other types of gyroscopes [88, 92].

REFERENCES

[1]   Sagnac, G., "L'ether lumineux demontre par l'effet du vent relatif d'ether dans un interferometre en rotation uniforme" ["The luminiferous aether demonstrated by the effect of the wind relative to the aether in a uniformly rotating interferometer"], *Comptes Rendus* 157, 708-710 (1913)

[2]   Malykin, G.B., "The Sagnac effect: correct and incorrect explanations," *Phys. Usp.* 43, 1229 (2000).

[3]   Лукьянов, Д.П., Распопов, В.Я., Филатов, Ю.В., *"Прикладная теория гироскопов,"* ["*The applied theory of gyros.*"] СПб: ГНЦ РФ ОАО "Концерн "ЦНИИ "Электроприбор", 316 (2015)

[4] Culshaw, B., "The optical fibre Sagnac interferometer: an overview of its principles and applications," *Meas. Sci. Technol.* 17, R1–R16 (2006).

[5] Li, Jiang, Myoung-Gyun Suh, and Kerry Vahala, "Microresonator Brillouin gyroscope," *Optica* 4, 346-348 (2017).

[6] Zhang, Liang, Yanping Xu, Ping Lu, Stephen Mihailov, Liang Chen, and Xiaoyi Bao, "Multi-Wavelength Brillouin Random Fiber Laser via Distributed Feedback From a Random Fiber Grating," *J. Lightwave Technol.* 36, 2122-2128 (2018).

[7] Liu, K., W. Zhang, W. Chen, K. Li, F. Dai, F. Cui, X. Wu, et al., "The development of micro-gyroscope technology," *J. Micromech. Microeng.* 19, 113001 (2009).

[8] Sorel, M., P.J.R. Laybourn, A. Scirè, S. Balle, G. Guiliani, R. Miglierina, and S. Donati, "Alternate oscillations in semiconductor ring lasers," *Opt. Lett.* 27, 1992–1994 (2002).

[9] Khandelwal, Arpit, Azeemuddin Syed, and Jagannath Nayak, "Performance Evaluation of Integrated Semiconductor Ring Laser Gyroscope," *J. Lightwave Technol.* 35, 3555-3561 (2017).

[10] Venediktov, V.Yu., Yu.V. Filatov, E.V. Shalymov, "Passive ring resonator micro-optical gyroscopes," *Quantum Electron.* 46(5), 437-446 (2016).

[11] Venediktov, V.Yu., Yu.V. Filatov, E.V. Shalymov, "On the possibility of using the phase characteristic of a ring interferometer in microoptical gyroscopes," *Quantum Electron.* 44(12), 1145-1150 (2014).

[12] Feng, Lishuang, Junjie Wang, Yinzhou Zhi, Yichuang Tang, Qiwei Wang, Haicheng Li, and Wei Wang, "Transmissive resonator optic gyro based on silica waveguide ring resonator," *Opt. Express* 22, 27565-27575 (2014).

[13] Mao, Hui, Huilian Ma, and Zhonghe Jin, "Polarization maintaining silica waveguide resonator optic gyro using double phase modulation technique," *Opt. Express* 19, 4632-4643 (2011).

[14] Ma, Huilian, Xuehui Li, Guhong Zhang, and Zhonghe Jin, "Reduction of optical Kerr-effect induced error in a resonant micro-

optic gyro by light-intensity feedback technique," *Appl. Opt.* 53, 3465-3472 (2014).

[15] Wang, Qiwei, Lishuang Feng, Hui Li, Xiao Wang, Yongze Jia, and Danni Liu, "Enhanced differential detection technique for the resonator integrated optic gyro," *Opt. Lett.* 43, 2941-2944 (2018).

[16] Ciminelli, C., F. Peluso, M.N. Armenise, "A new integrated optical angular velocity sensor," *Proc. of SPIE.* 5728, 93-100 (2005).

[17] Hsiao Hsien-kai, and K.A. Winick, "Planar glass waveguide ring resonators with gain," *Opt. Express* 15, 17783-17797 (2007).

[18] Rabus, D.G., Hamacher, M., Troppenz, U., Heidrich, H., "Optical filters based on ring resonators with integrated semiconductor optical amplifiers in GaInAsP–InP," *IEEE J. Sel. Top. Quantum Electron.* 8, 1405-1411 (2002).

[19] Rasoloniaina, A., S. Trebaol, V. Huet, E. Le Cren, G. Nunzi Conti, H. Serier-Brault, M. Mortier, Y. Dumeige, and P. Féron, "High-gain wavelength-selective amplification and cavity ring down spectroscopy in a fluoride glass erbium-doped microsphere," *Opt. Lett.* 37, 4735-4737 (2012).

[20] Zhang, R., Wang, J. Qiu, B., Wang, K., Wang, S., "Design of High-Q Compact Passive Ring Resonators via Incorporating a Loss-Compensated Structure for High Performance Angular Velocity Sensing in Monolithic Integrated-Optical-Gyroscopes," *IEEE Sensors Journal.* 17, 84-90 (2017).

[21] Rasoloniaina, A., V. Huet, T.K.N. Nguyen, E. Le Cren, Michel Mortier, L. Michely, Y. Dumeige, P. Féron, "Controling the coupling properties of active ultrahigh-Q WGM microcavities from undercoupling to selective amplification," *Scientific Reports* 4, 4023 (2014).

[22] Filatov, Yuri V., Alina V. Gorelaya, Egor V. Shalymov, and Vladimir Yu. Venediktov "Optical gyros operating using the phase characteristic of the ring confocal resonator," *Proc. SPIE* 10821, 108210B (2018).

[23] Боронахин, А.М., Лукьянов, Д.П., Филатов, Ю.В., *"Оптические и микромеханические инерциальные*

приборы" ["*Optical and micromechanical inertial devices*"]. СПб.: Элмор, 400 (2008)

[24] Srinivasan, Sudharsanan, Renan Moreira, Daniel Blumenthal, and John E. Bowers, "Design of integrated hybrid silicon waveguide optical gyroscope," *Opt. Express* 22, 24988-24993 (2014).

[25] Gundavarapu, Sarat, Michael Belt, Taran Huffman, Minh A. Tran, Tin Komljenovic, John E. Bowers, Daniel J. Blumenthal, "Interferometric Optical Gyroscope Based on an Integrated Si3N4 Low-Loss Waveguide Coil," *Journal of Lightwave Technology* 36(4), 1185-1191 (2017).

[26] Huffman, T., Davenport, M., Belt, M., Bowers, J., Blumenthal, D., "Ultra-Low Loss Large Area Waveguide Coils for Integrated Optical Gyroscopes," *IEEE Photonics Technology Letters* 29, 185-188 (2017).

[27] Beibei Wu, Yu Yu, Jiabi Xiong, Xinliang Zhang, "Silicon Integrated Interferometric Optical Gyroscope," *Scientific Reports* 8, 8766 (2018).

[28] Wei, W., Junlei, X., Yuxin X., "Research on integrated optical gyroscope," *Proc. of IEEE 2nd International Symposium on System and Control in Aerospace and Astronautics (ISAACAA)*, 10478339 (2008).

[29] Fan, Geng, Sun Fengyu, "Research on a new type of re-entry IFOG," *Infrared and Laser Engineering* 33(1), 10-13 (2004).

[30] Mao, Xianhui, Qian Tian, Enyao Zhang, Liqun Sun, Yunhe Teng, Jinjiang Liu, "Research on multiple-integrated-optic-chip of re-entrant fiber optic gyro," *Proc. SPIE* 5826, Opto-Ireland 2005: Optical Sensing and Spectroscopy, (2005).

[31] Zhang, Y.S., Ding H.G., "Investigation of system configuration for micro optic gyros," Herald of the Bauman Moscow State Technical University, *Instrument Engineering* 4, 109-117 (2005).

[32] Adar, R.R., Serbin, M.R., Mizrahi V., "Less than 1 dB per meter propagation loss of silica waveguides measured using a ring resonator," *J. Lightwave Technol.* 12(8), 1369-1372 (1994).

[33] Yu, H., Zhang, C., Feng, L., et al. "SiO$_2$ Waveguide resonator used in an integrated optical gyroscope," *Phys. Lett.* 26(5), 054210 (2009).

[34] Zhang, Jianjie, Hanzhao Li, Huilian Ma, and Zhonghe Jin "High finesse silica waveguide ring resonators for resonant micro-optic gyroscopes," *Proc. SPIE 10323, 25th International Conference on Optical Fiber Sensors*, 1032319 (2017).

[35] Tien, Ming-Chun, Jared F. Bauters, Martijn J.R. Heck, Daryl T. Spencer, Daniel J. Blumenthal, and John E. Bowers, "Ultra-high quality factor planar Si$_3$N$_4$ ring resonators on Si substrates," *Opt. Express* 19, 13551-13556 (2011).

[36] Spencer, D.T., Tang, Y., Bauters J.F., et al. "Integrated Si3N4/SiO2 ultra high Q ring resonators," *Proc. to the IEEE Photonics Conference* (ICP) 13149871. 141–142 (2012).

[37] Bauters, Jared F., Martijn J.R. Heck, Demis D. John, Jonathon S. Barton, Christiaan M. Bruinink, Arne Leinse, René G. Heideman, Daniel J. Blumenthal, and John E. Bowers, "Planar waveguides with less than 0.1 dB/m propagation loss fabricated with wafer bonding," *Opt. Express* 19, 24090-24101 (2011).

[38] Biberman, A., Shaw, M.J., Timurdogan E., et al. "Ultralow-loss silicon ring resonators," *Optic Lett.* 37(20), 4236-4238 (2012).

[39] Vannahme, C., H. Suche, S. Reza et al., "Integrated optical Ti:LiNbO3 ring resonator for rotation rate sensing," *Proc. to the 13th European Conference on Integrated Optics, WE1* (2007).

[40] Zhang, Mian, Cheng Wang, Rebecca Cheng, Amirhassan Shams-Ansari, and Marko Lončar, "Monolithic ultra-high-Q lithium niobate microring resonator," *Optica* 4, 1536-1537 (2017).

[41] Tolstikhin, V.I., A. Densmore, Y. Logvin, K. Pimenov, F. Wu, and S. Laframboise, "44-channel optical power monitor based on an echelle grating demultiplexer and a waveguide photodetector array monolithically integrated on an InP substrate," *Proceedings of the Optical Fiber Communication Conference* OFC 2003, PD37 (2003).

[42] Nicholes, S., M.L. Mašanovic, B. Jevremovic, E. Lively, L. Coldren, and D.J. Blumenthal, "The World's First InP 8x8 Monolithic

Tunable Optical Router (MOTOR) Operating at 40 Gbps Line Rate per Port," *Optical Fiber Communication Conference OFC* 2009, PDPB1 (2009).

[43] Soares, F.M., N.K. Fontaine, R.P. Scott, J.H. Baek, X. Zhou, T. Su, S. Cheung, Y. Wang, C. Junesand, S. Lourdudoss, K.Y. Liou, R.A. Hamm, W. Wang, B. Patel, L.A. Gruezke, W.T. Tsang, J.P. Heritage, and S.J.B. Yoo, "Monolithic InP 100-Channel × 10-GHz Device for Optical Arbitrary Waveform Generation," *IEEE Photon. J.* 3(6), 975-985 (2011).

[44] Ciminelli, Caterina, Francesco Dell'Olio, Mario N. Armenise, Francisco M. Soares, and Wolfgang Passenberg, "High performance InP ring resonator for new generation monolithically integrated optical gyroscopes," *Opt. Express* 21, 556-564 (2013).

[45] Eldada, L., L.W. Shacklette, "Advances in polymer integrated optics," *IEEE Journal on Selected Topics in Quantum Electronics* 6(1), 54-68 (2000).

[46] Chen, J.-G., T. Zhang, J.-S. Zhu et al. "Low-loss planar optical waveguides fabricated from polycarbonate," *Polymer Engineering and Science* 49(10), 2015-2019 (2009).

[47] Qian, Guang, Jie Tang, Xiao-Yang Zhang, Ruo-Zhou Li, Yu Lu, and Tong Zhang, "Low-Loss Polymer-Based Ring Resonator for Resonant Integrated Optical Gyroscopes," *Journal of Nanomaterials* 2014, 146510 (2014).

[48] Tu, X., Chen, S.-L., Song, C., Huang, T., Guo, L., "Ultrahigh Q polymer microring resonators for biosensing applications," *IEEE Photonics Journal* (2019). DOI: 10.1109/JPHOT.2019.2899666.

[49] Terrel, Matthew A., Michel J.F. Digonnet, and Shanhui Fan "Coupled resonator gyroscopes: what works and what does not," *Proc. SPIE* 7612, 76120B (2010).

[50] Terrel, M.A., M.J.F. Digonnet, S. Fan, "Performance limitations of a coupled resonant optical waveguide gyroscope," *J. Lightwave Technol.* 27(1), 47-54 (2009).

[51] Terrel, M.A., M.J.F. Digonnet, S. Fan, "Performance comparison of slow-light coupled-resonator optical gyroscopes," *Laser Photonics Rev* 3(5), 452-464 (2009).

[52] Kiarash Zamani Aghaie, Pierre-Baptiste Vigneron, and Michel J.F. Digonnet "Rotation sensitivity analysis of a two-dimensional array of coupled resonators," *Proc. SPIE* 9378, 93781P (2015).

[53] Toland John R.E., Search, Christopher P. "Sagnac gyroscope using a two-dimensional array of coupled optical microresonators," *Applied Physics B* 114(3), 333-339 (2014).

[54] Kalantarov Dmitriy, and Christopher P. Search, "Effect of input–output coupling on the sensitivity of coupled resonator optical waveguide gyroscopes," *J. Opt. Soc. Am.* B 30, 377-381 (2013).

[55] Kalantarov Dmitriy, and Christopher P. Search, "Effect of resonator losses on the sensitivity of coupled resonator optical waveguide gyroscopes," *Opt. Lett.* 39, 985-988 (2014)

[56] Florio, F., D. Kalantarov, C.P. Search, "Effect of Static Disorder on Sensitivity of Coupled Resonator Optical Waveguide Gyroscopes," *J. Lightwave Technol.* 32(21), 4020-4028 (2014).

[57] Scheuer, J., A. Yariv, "Sagnac effect in coupled-resonator slowlight waveguide structures," *Phys. Rev. Lett.* 96, 053901 (2006).

[58] Toland, John R.E., Zachary A. Kaston, Christopher Sorrentino, and Christopher P. Search, "Chirped area coupled resonator optical waveguide gyroscope," *Opt. Lett.* 36, 1221-1223 (2011).

[59] Peng, C., Z. Li, A. Xu, "Optical gyroscope based on a coupled resonator with the all-optical analogous property of electromagnetically induced transparency," *Optics Express* 15(7), 3864-3875(2007).

[60] Sorrentino, Christopher, John R.E. Toland, and Christopher P. Search, "Ultra-sensitive chip scale Sagnac gyroscope based on periodically modulated coupling of a coupled resonator optical waveguide," *Opt. Express* 20, 354-363 (2012).

[61] Novitski, Roman, Ben, Z. Steinberg, and Jacob Scheuer, "Finite-difference time-domain study of modulated and disordered coupled

resonator optical waveguide rotation sensors," *Opt. Express* 22, 23153-23163 (2014).

[62] Zhang, Y., N. Wang, H. Tian et al. "A high sensitivity optical gyroscope based on slow light in coupled-resonator-induced transparency," *Phys. Lett. A.* 372(36), 5848-5852 (2008).

[63] Tian, He, Yundong Zhang, Xuenan Zhang, Hao Wu, and Ping Yuan, "Rotation sensing based on a side-coupled spaced sequence of resonators," *Opt. Express* 19, 9185-9191 (2011).

[64] Matsko, A.B., A.A. Savchenkov, V.S. Ilchenko et al. "Optical gyroscope with whispering gallery mode optical cavities," *Opt. Commun.* 233(1), 107-112 (2004).

[65] Ciminelli, C.C., F. Dell'Olio, M.N. Armenise, "High-Q spiral resonator for optical gyroscope applications: numerical and experimental investigation," *IEEE Photon. J.* 4(5), 1844-1854 (2012).

[66] Wu, Beibei, Yu Yu, Jiabi Xiong, Xinliang Zhang, "Silicon Integrated Interferometric Optical Gyroscope," *Scientific Reports* 8, 8766 (2018).

[67] Ciminelli, Caterina, Domenico D'Agostino, Giuseppe Carnicella, Francesco Dell'Olio, Donato Conteduca, Huub P.M.M. Ambrosius, Meint K. Smit, Mario N. Armenise, "A High-Q InP Resonant Angular Velocity Sensor for a Monolithically Integrated Optical Gyroscope," *IEEE Photonics Journal* 8(1), 7352311 (2016).

[68] Sumetsky, M., Y. Dulashko, J.M. Fini et al. "The microfiber loop resonator: theory, experiment, and application," *J. Lightwave Technol.* 24(1), 242-250 (2006).

[69] Stokes, L.F., M. Chodorow, and H.J. Shaw, "All-single-mode fiber resonator," *Opt. Lett.* 7, 288-290 (1982).

[70] Belal, M., Z. Song, Y. Jung, G. Brambilla, and T.P. Newson, "Optical fiber microwire current sensor," *Opt. Lett.* 35, 3045-3047 (2010).

[71] Xu, F., V. Pruneri, V. Finazzi et al. "An embedded optical nanowire loop resonator refractometric sensor," *Optics Express* 16(2), 1062-1067 (2008).

[72] Ma, Huilian, Zhen Chen, Zhihuai Yang, Xuhui Yu, and Zhonghe Jin, "Polarization-induced noise in resonator fiber optic gyro," *Appl. Opt.* 51, 6708-6717 (2012).

[73] Strandjord, Lee K., Tiequn Qiu, Jianfeng Wu, Thomas Ohnstein, Glen A. Sanders, "Resonator fiber optic gyro progress including observation of navigation grade angle random walk," *Proc. SPIE* 8421, 842109 (2012).

[74] Strandjord, L.K., T. Qiu, M. Salit, C. Narayanan, M. Smiciklas, J. Wu, and G.A. Sanders, "Improved Bias Performance in Resonator Fiber Optic Gyros using a Novel Modulation Method for Error Suppression," in *26th International Conference on Optical Fiber Sensors*, ThD3 (2018).

[75] Joannopoulos, J.D., S.G. Johnson, J.N. Winn et al. *Photonic Crystals: Molding the Flow of Light*, Princeton University Press, 302 (2008).

[76] Ciminelli, C., E.C. Campanella, M.N. Armenise, "*Optical rotation sensor as well as method of manufacturing an optical rotation sensor*," European Patent 2917691 (2013).

[77] Goldring, Damian, Uriel Levy, and David Mendlovic, "Highly dispersive micro-ring resonator based on one dimensional photonic crystal waveguide design and analysis," *Opt. Express* 15, 3156-3168 (2007).

[78] Ciminelli, Caterina, F. Innone, G. Brunetti, Donato Conteduca, Francesco Dell'Olio, Teresa Tatoli, Mario Nicola Armenise, "Rigorous model for the design of ultra-high Q-factor resonant cavities," *Proc. 18th International Conference on Transparent Optical Networks (ICTON)*, 1-4 (2016).

[79] Yong Zhang, Cheng Zeng, Danping Li, Ge Gao, Zengzhi Huang, Jinzhong Yu, and Jinsong Xia, "High-quality-factor photonic crystal ring resonator," *Opt. Lett.* 39, 1282-1285 (2014).

[80] Steinberg, B.Z., A. Boag, "Splitting of microcavity degenerate modes in rotating photonic crystals - the miniature optical gyroscopes," *J. Opt. Soc. Am. B.* 24(1), 142-151 (2007).

[81] Mohammadi, Masoud, Saeed Olyaee, Mahmood Seifouri, "Passive Integrated Optical Gyroscope Based on Photonic Crystal Ring Resonator for Angular Velocity Sensing," *Silicon,* 1-8 (2018).

[82] Filatov, Y.V., Sevryugin, A.A., Shalymov, E.V., Venediktov, V.Y., "Frequency properties of the confocal ring cavity," *Optical Engineering* 54(4), 044107 (2015).

[83] Kuryatov, V.N., Sokolov, A.L., "Polarisation losses in a ring prism cavity," *Quantum Electronics* 30(2), 125-127 (2000).

[84] Oraevsky, A.N. "Whispering-gallery waves," *Quantum Electron.* 32(5), 377-400 (2002).

[85] Vollmer, F., L. Yang, "Label-free detection with high-Q microcavities: a review of biosensing mechanisms for integrated devices" *Nanophotonics* 1(3), 267-291 (2012).

[86] Yu, V., Venediktov, A.S. Kukaev, Yu. V. Filatov, E.V. Shalymov, "Modelling of rotation-induced frequency shifts in whispering gallery modes", *Quantum Electron.* 48(2), 95-104 (2018).

[87] Liang W., Ilchenko V., Eliyahu D., Dale E., Savchenkov A., Matsko A.B., Maleki L. "Whispering gallery mode optical gyroscope," *Proc. IEEE International Symposium on Inertial Sensors and Systems* 7435552, 89-92 (2016).

[88] Liang, Wei, Vladimir S. Ilchenko, Anatoliy A. Savchenkov, Elijah Dale, Danny Eliyahu, Andrey B. Matsko, and Lute Maleki, "Resonant microphotonic gyroscope," *Optica* 4, 114-117 (2017).

[89] Savchenkov, Anatoliy A., Andrey B. Matsko, Vladimir S. Ilchenko, and Lute Maleki, "Optical resonators with ten million finesse," *Opt. Express* 15, 6768-6773 (2007).

[90] Guan, G., S. Arnold, M.V. Otugen, "Temperature Measurements Using a Microoptical Sensor Based on Whispering Gallery Modes," *AIAA Journal* 44(10), 2385-2389 (2006).

[91] Ioppolo, T., V. Ötügen, D. Fourguette et al. "Effect of acceleration on the morphology-dependent optical resonances of spherical resonators," *Opt. Soc. Am. B.* 28(2), 225-227 (2011).

[92] Dmitrieva, A.D., Yu.V. Filatov, E.V. Shalymov al. "Application of optical whispering gallery mode resonators for rotation sensing," *Proc. of SPIE* 9899, 98991L (2016).

INDEX

ADVANCES IN AEROSPACE SCIENCE AND TECHNOLOGY: PART II

EDITORS: Parvathy Rajendran and Mohd Zulkifly Abdullah

SERIES: Mechanical Engineering Theory and Applications

BOOK DESCRIPTION: Aerospace science and technology have made remarkable progress in the last century. This book presents state-of-the-art and current developments and applications in aerospace.

HARDCOVER ISBN: 978-1-53615-689-8
RETAIL PRICE: $160

MECHANICAL DESIGN, MATERIALS AND MANUFACTURING

EDITOR: Sandip A. Kale

SERIES: Mechanical Engineering Theory and Applications

BOOK DESCRIPTION: Though the developments in the field of electronics and digital industries are significant, the importance of the basic mechanical industry remains always on the top side. The purpose of this book is to present some advanced research studies on mechanical design, materials and manufacturing.

HARDCOVER ISBN: 978-1-53614-791-9
RETAIL PRICE: $230